ALEX
THROUGH THE
LOOKING - GLASS

BY THE SAME AUTHOR

Futebol
Alex's Adventures in Numberland

ALEX THROUGH THE LOOKING-GLASS

HOW NUMBERS REFLECT LIFE AND LIFE REFLECTS NUMBERS

ALEX BELLOS

Illustrations by The Surreal McCoy

BLOOMSBURY

LONDON · NEW DELHI · NEW YORK · SYDNEY

For Nat

First published in Great Britain in 2014
Copyright © 2014 by Alex Bellos
Cartoons © 2014 by The Surreal McCoy
Mathematical diagrams © 2014 by Simon Lindo

The moral right of the author has been asserted

Bloomsbury Publishing Plc, 50 Bedford Square, London WC1B 3DP

A CIP catalogue record for this book is available from the British Library

Hardback ISBN 978 1 4088 1777 3
Trade paperback ISBN 978 1 4088 5098 5

10 9 8 7 6 5 4 3 2 1

Designed by Libanus Press

Printed and bound in Great Britain by CPI Group (UK) Ltd, Croydon CR0 4YY

alexbellos.com
bloomsbury.com/alexbellos

Plate Section Picture Credits:
p. 1 (top) © iStock.com/dolgachov, as used in *Are Numbers Gendered?*; p. 1 (bottom),
p. 15 (top) © Alex Bellos; p. 2 © Bob Sacha/Corbis; p. 3 (top) © British Library/Science
Photo Library; p. 3 (bottom) © Rob Woodall; p. 4 (left) Laurents/House & Garden
© Condé Nast; pp. 4–5 (top and right), p. 15 (bottom) © Getty Images; p. 6 (top)
© Science Museum/Science & Society Picture Library; p. 6 (bottom) © iStock.com/
real444; p. 7 Paramount/The Kobal Collection; pp. 8–9 © Science Museum/Science
& Society Picture Library; p. 10 © Fundació Institut Amatller d'Art Hispànic, Arxiu
Mas; p. 11 (top) © Foster + Partners; p. 11 (bottom) © iStock.com/lucylui; pp. 12–13
© Daniel White; p. 14 © Eric Le Roux/Université Claude Bernard Lyon 1; p. 16 (top)
© Association Nicolas Bourbaki; p. 16 (bottom) Photograph by Nicholas Metropolis,
courtesy of Claire Ulam.

Contents

CHAPTER TEN   257

Cell Mates

In which the author voyages into the realm of the cellular automaton. He explores the meaning of Life and speaks to the man looking for universes in his basement.

Introduction

Mathematics is a joke.

I'm not being funny.

You need to 'get' a joke just like you need to 'get' maths.

The mental process is the same.

Think about it. Jokes are stories with a set-up and a punch line. You follow them carefully until the payoff, which makes you smile.

A piece of maths is also a story with a set-up and a punch line. It's a different type of story, of course, in which the protagonists are numbers, shapes, symbols and patterns. We'd usually call a mathematical story a 'proof', and the punch line a 'theorem'.

You follow the proof until you reach the payoff. Whoosh! You get it! Neurons go wild! A rush of intellectual satisfaction justifies the initial confusion, and you smile.

The *ha-ha!* in the case of a joke and the *aha!* in the case of maths describe the same experience, and this is one of the reasons why understanding mathematics can be so enjoyable and addictive.

Like the funniest punch lines, the finest theorems reveal something you are not expecting. They present a new idea, a new perspective. With jokes, you laugh. With maths, you gasp in awe. It was precisely this element of surprise that made me fall in love with maths as a child. No other subject so consistently challenged my preconceptions.

The aim of this book is to surprise you, too. In it, I embark on a tour of my favourite mathematical concepts, and explore their presence in our lives. I want you to appreciate the beauty, utility and playfulness of logical thought.

My previous book, *Alex's Adventures in Numberland*, was a journey into mathematical abstraction. This time I come down to earth: my concern is as much the real world, reflected in the mirror of maths, as it is the abstract one, inspired by our physical experiences.

Firstly, I put humans on the couch. What are the feelings we have for numbers, and what triggers these feelings? Then I put numbers on the couch, individually and as a group. Each number has its own issues. When we engage with them *en masse*, however, we see fascinating behaviour: they conduct themselves like a well-organized crowd.

We depend on numbers to make sense of the world, and have done so ever since we started to count. In fact, perhaps the most surprising feature of mathematics is how extraordinarily successful it has been, and continues to be, in enabling us to understand our surroundings. Civilization has progressed as far as it has thanks to discoveries about simple shapes like circles and triangles, expressed pictorially at first, and later in the vernacular of equations.

Maths, I would argue, is the most impressive and longest-running collective enterprise in human history. In the following pages I follow the torch of discovery from the Pyramids to Mount Everest, from Prague to Guangzhou, and from the Victorian drawing room to a digital universe of self-replicating creatures. We will meet swashbuckling intellects, including familiar names from antiquity and less familiar names from the present day. Our cast includes a cravat-wearing celebrity in India, a gun-toting private investigator in the United States, a member of a secret society in France, and a spaceship engineer who lives near my London flat.

As we roam across physical and abstract worlds, we will probe well-known concepts, like pi and negative numbers, and encounter more enigmatic ones, which will become our confidantes. We will marvel at concrete applications of mathematical ideas, including some that actually are made of concrete.

You don't need to be a maths whizz to read this book. It's aimed at the general reader. Each chapter introduces a new mathematical concept, and assumes no previous knowledge. Inevitably, however, some concepts are more stretching than others. The level sometimes reaches that of an undergraduate degree, and, depending on your mathematical proficiency, there may be moments of bewilderment.

In these cases, skip to the beginning of the next chapter, where I reset the level to elementary. The material might make you feel a bit dizzy at first, especially if it is new to you, but that's the point. I want you to see life differently. Sometimes the *aha!* takes time.

If all this sounds a bit serious, it isn't. The emphasis on surprise has made maths the most playful of all intellectual disciplines. Numbers have always been toys, as much as they have been tools.

Not only does maths help you understand the world better, it helps you enjoy it more, too.

Alex Bellos
January 2014

Every Number Tells a Story

Jerry Newport asked me to pick a four-digit number.

'2761,' I said.

'That's 11 × 251,' he replied, reciting the numbers in one continuous, unhesitant flow.

'2762. That's 2 × 1381.

'2763. That's 3 × 3 × 307.

'2764. That's 2 × 2 × 691.'

Jerry is a retired taxi driver from Tucson, Arizona with Asperger syndrome. He has a ruddy complexion and small blue eyes, his large forehead sliced by a diagonal comb of dark blond hair. He likes birds as well as numbers, and when we met he was wearing a flowery red shirt with a parrot on it. We were sitting in his living room, together with a cockatoo, a dove, three parakeets and two cockatiels, which were also listening to, and occasionally repeating, our conversation.

As soon as Jerry sees a big number, he divides it up into prime numbers, which are those numbers – 2, 3, 5, 7, 11 … – that can only be divided by themselves and 1. This habit made his former job driving cabs particularly enjoyable, since there was always a number on the licence plate in front of him. When he lived in Santa Monica, where licence numbers were four and five digits long, he would often visit the four-storey car park of his local mall and not leave until he had worked through every plate.

In Tucson, however, car numbers are only three digits long. He barely glances at them now.

'If the number is more than four digits I'll start to pay attention to it. If it's four digits or less, it's roadkill. It is!' he remonstrated. 'Come on! Show me something new!'

Asperger's is a psychological disorder in which social awkwardness can coexist with extreme abilities, such as, in Jerry's case, an extraordinary talent for mental arithmetic. In 2010 he competed at the Mental Calculation World Cup in Germany having done no preparation. He won the overall title of Most Versatile Calculator, the only contestant to score full marks in the category where 19 five-digit numbers have to be decomposed into their constituent primes in ten minutes. No one else got even close.

Jerry's system for breaking down large numbers is to sieve out the prime numbers in ascending order, extracting a 2 if the number is even, extracting a 3 if it divides by three, a 5 if it divides by five, and so on.

He raised his voice to a yell: 'Oh yeah, we're sievin', baby!' He started moving his body around: 'We're on stage. Throw those numbers out, crowd, and we'll sieve 'em for ya! Yeah! Jerry and the Sievers!'

'I've got a pair of sievers,' interrupted his wife, Mary, who was sitting on the sofa next to us. Mary, a musician and former Star Trek extra, also has Asperger's, which is much less common in women than it is in men. A marriage between two people with Asperger's is very rare, and their unconventional romance was turned into the 2005 Hollywood movie *Mozart and the Whale*.

Sometimes Jerry cannot extract any primes at all from a large number, which means the number is itself prime. When this happens it gives him a thrill: 'If it's a prime number I've never found before, it's kinda like if you were looking for rocks, and you've found a new rock. Something like a diamond you can take home and put on your shelf.'

He paused. 'A new prime number – it's like having a new friend.'

The earliest words and symbols used for numbers date from about 5000 years ago in Sumer, a region in what is now Iraq. The Sumerians did not look far when coming up with names. The word for one, *ges,* also meant man, or erect phallus. The word for two, *min,* also meant woman, symbolic of the male being primary and the woman his complement, or perhaps describing a penis and a pair of breasts.

Initially, numbers served a practical purpose, like counting sheep and calculating taxes. Yet numbers also revealed abstract patterns, which made them objects of deep contemplation. Perhaps the earliest mathematical discovery was that numbers come in two types, even and odd: those that can be halved cleanly, such as 2, 4 and 6, and those that cannot, such as 1, 3 and 5. The Greek teacher Pythagoras, who lived in the sixth century BCE, echoed the Sumerian association of one with man and two with woman by proclaiming odd numbers masculine and even numbers feminine. Resistance to splitting in two, he argued, embodied strength, while susceptibility to splitting in two was a weakness. He gave a further arithmetical justification: odd was master over even, just as man is master over woman, because when you add an odd number to an even number, the answer remains odd.

Pythagoras is most famous for his theorem about triangles, which we will come to later. But his belief about number gender has dominated Western thought for more than two thousand years. Christianity embraced it within its creation myth: God created Adam first, and Eve second. One signifies unity, and two is the 'sin which deviates from the First Good'. For the medieval Church, odd numbers were stronger, better, more godly and luckier than the evens, and by Shakespeare's time, metaphysical beliefs about odd numbers were common: 'They say there is divinity in odd numbers, either in nativity, chance or death,' Falstaff declares in *The Merry Wives of Windsor*. These superstitions remain. Mystical numbers still tend to be odd, notably the 'magic' three, the 'lucky' seven, and the 'unlucky' thirteen.

Shakespeare is also responsible for the modern meaning of 'odd'. Originally, the word only had a numerical sense. It was used in phrases such as 'odd man out', the unpaired member of a group of three. But in *Love's Labour's Lost*, the farcical Spaniard Don Adriano de Armado is described as 'too picked, too spruce, too affected, too odd, as it were'. Having one left over when divided by two has meant peculiar ever since.

It is human nature to be sensitive to numerical patterns. These patterns provoke subjective responses, sometimes extreme ones, as we saw with Jerry Newport, but also more generally, leading to deeply held cultural associations. Oriental philosophy is based on an appreciation of the dualities in nature, symbolized by *yin* and *yang*, literally 'shadow' and 'light'. *Yin* is associated with passivity, femininity, the moon, misfortune and even numbers, and *yang* with their complements: aggressiveness, masculinity, the sun, good fortune and odd numbers. Again, we see a historic link between luck and oddness, and this link is especially strong in Japan, where, for example, it is customary to give three, five or seven items as a gift. Never four or six. When giving cash to newlyweds, the amounts ¥30,000, ¥50,000 and ¥100,000 are preferred, although ¥20,000 is acceptable, in which case the recommendation is to 'odd things out' by dividing the value into one ¥10,000 and two ¥5,000 bills. The aesthetics of odd numbers also underpins the Japanese classical art of flower arranging, *ikebana*, which uses only odd numbers of items, an influence of the Buddhist belief that asymmetry reflects nature. A meal of Japanese haute cuisine, *kaiseki*, always comprises an odd number of dishes, and, just so kids get the message early on, the annual celebration of youthful good health is called the Seven-Five-Three festival, in which only children who are three, five and seven years old take part. The Japanese taste for odd numbers is so ingrained, wrote Professor Yutaka Nishiyama of the Osaka University of Economics, that when the government released a ¥2,000 note in 2000 no one ever used it.

(Number superstitions are stronger in East Asian countries than they are in the West. These countries also score higher in international tests of numeracy, indicating that strong mystical beliefs about numbers are not necessarily an impediment to learning arithmetical skills. Superstitions, in fact, may encourage a respect for numbers, and an intimacy and playfulness with them – just like mathematics does. The most widely held Asian number belief is based on a pun. Because the words for 'four' in Japanese, Cantonese, Mandarin and Korean (*shi, sei, si, sa*) sound the same as the words in those languages for death, the number four is avoided when at all possible. Hotels across the region often do not have fourth floors, aeroplanes often do not have fourth rows and companies often do not release product lines with a 4 in the name. Indeed, the association of four with death is so strongly held that it has become a self-fulfilling prophecy: US records show that among Japanese and Chinese Americans, fatal heart attacks surge on the fourth of each month. Eight, however, is lucky, because the word in Chinese sounds like 'prosperity'. The number 8 appears disproportionately often in retail prices in Chinese newspaper adverts. Two deaths equal a prosperous life.)

In India too, odd numbers are seen as the most auspicious type of number. Is there any reason why in both the East and the West odd numbers are imbued with more spiritual significance than even ones? It may be related to the fact that our brains spend longer processing odd numbers than they do evens, a phenomenon discovered by the psychologist Terence Hines of Pace University, which he calls the 'odd effect'. In one experiment Hines displayed pairs of digits on a screen. Either the digits were both odd, such as 35, both even, such as 64, or one even and one odd, such as 27. He told respondents to press a button only when the digits were even-even or odd-odd. On average it took respondents about 20 per cent longer to press the button when both digits were odd, and they made more mistakes. At first Hines did not believe his results, thinking there must be a flaw in his testing procedure, but the phenomenon showed up clearly in follow-up research. We feel differently towards odd numbers not just because of age-old cultural beliefs but because we *think* differently about them. They are literally more thought-provoking.

There is a linguistic clue to the odd effect, invisible to speakers of English, which is the only major European language to have unrelated words for odd and even. In French, German and Russian, for example, the words for even and odd are 'even' and 'not-even': *pair/impair, gerade/ungerade,* and *chyotny/nyechyotny.* Evenness is an idea that precedes oddness. It is a simpler concept, easier to understand.

The cognitive gap between odd and even numbers has been the subject of other studies. James Wilkie and Galen Bodenhausen of Northwestern University decided to investigate whether there was any psychological basis to the ancient belief that odds are male and evens are female. They showed respondents randomly assigned pictures of the faces of young babies, each next to a three-digit number that was either odd-odd-odd or even-even-even, and asked them to guess the baby's sex. This experiment sounds absurd, and it would have been forgotten had it not achieved a striking result: the choice of number had a significant effect. Respondents were about 10 per cent more likely to say that a baby paired with odd numbers was a boy, than if the same baby was paired with even numbers. Wilkie and Bodenhausen concluded that the Pythagoreans, medieval Christians and Taoists were right. The ancient, cross-cultural belief that odds are associated with maleness and evens with femaleness was supported by the data. 'It may indeed be a universal human tendency to project gendered meanings onto numbers,' they wrote. They were unable to explain, however, why odd is masculine and even is feminine, rather than vice versa.

Culture, language and psychology play a role in the way we understand mathematical patterns, as we have seen here with odd numbers, and as we will see shortly with other numerical cues. Numbers have a fixed mathematical meaning – they are abstract entities signifying quantity and order – yet they also tell other stories.

The influential German theologian Hugh of Saint Victor (1096–1141) provided an early guide to numbers: ten represents 'rectitude in faith', nine, coming before ten, 'defect among perfection', and eleven, coming afterwards, 'transgression outside of measure'. If Hugh were alive today he would undoubtedly find lucrative employment at The Semiotic Alliance, one of the world's leading semiotics agencies. I met its founder, Greg Rowland, in London. With a black and white T-shirt under his jacket, strong lines on his forehead and sharp eyes, he came across as a groovy university professor, although his habitat is not the library but the executive boardroom. Greg advises multinational companies on the symbolism of their brands, which involves the cultural associations of numbers. His clients include Unilever, Calvin Klein and KFC. The number eleven, for example, is an essential element of KFC's corporate mythology: its signature dish is fried chicken seasoned with Colonel Sanders' secret Original Recipe of eleven herbs and spices. 'This is the key mystical use of the number eleven in commercial culture,' said Greg. The number represents transgression, he added, in this case an extra ingredient, one beyond the ordinary. 'Eleven has just gone that one past ten. It has recognized that there is an order to things, and now it is exploring the distance beyond. Eleven is opening the door to the infinite, but it's not going too far. It is … bourgeois rebellion at its most finite!' I asked if Colonel Sanders was therefore no different from the rocker in *Spinal Tap* whose amp went up to 11 so it could be louder than amps labelled to 10. Greg laughed: 'Yes! But I actually believe it! I believe that 11 is more interesting than 10!'

The *Spinal Tap*-style extra 1, he added, is a common meme. A classic example is Levi's 501 jeans. 'This raises the expectation but doesn't overplay it. It's that extra little bit, and that is what Levi's is always doing, or in its glory days always did: adding an extra little button here, or a new piece of sewing there. It really was just a 1. Levi's is saying it's not just 500, it's one better than that, in the way that 502 – two better – doesn't work. There is that mystical added element, which stops it being as definable and reasonable as 500. For the big decimals it works best: the film *2001: A Space Odyssey*, the 101 drum machine, Room 101. It wasn't Room 100 – who'd be scared of that?'

Long before Levi's started making jeans, the significance of the extra 1 was an established part of Indian culture. *Shagun* is the tradition that gifts of money are always a round sum with one rupee added, such as Rs101, Rs501 or Rs100,001. Gift envelopes in wedding shops, for example, come with a one rupee coin glued to them, so you don't forget it. While there is no single explanation for the practice – some say the 1 is a blessing, others that it represents the beginning of a new cycle – it is accepted that the symbolic value of the extra 1 is as important as the monetary value of the notes inside.

Which brings me to an old family story. In the early twentieth century my grandfather was working on a new recipe for carbonated lemonade. He called it 4 Up. Consumers did not take to it, so he spent a few years developing it some more. His next launch, 5 Up, also failed. After another few years he released 6 Up, and guess what? It flopped, too. Grandad died, tragically, without knowing how close he came.

Yes, it's an old joke. But it contains a truth. In business, as in religion, a good number is fundamental. The number ten – 'rectitude in faith' – strengthens faith in the anti-acne cream Oxy 10: 'Ten is about balance, security, returning to the norm. It's the absolute decimal,' said Greg. 'There is no argument with 10, and that's what you want with spot things. You don't want Oxy 9, or Oxy 8 even. You certainly don't want Oxy 7 or 11 or 13 or 15. For a product like Oxy 10 you want certainty.' I asked him if he thought the all-purpose lubricant WD-40 would have been as successful if it had been called WD-41. 'WD-41 would not be reliable,' he insisted. 'WD-41 would have more stuff in it than you would want. It would have some extra bit in it, wouldn't it?' He thought aloud about other variants: 'WD-10 would have binary function. Either it did something or it didn't. But then it is not WD-400 or 4000 – you don't want to overdo it! WD-40 is not over-claiming. It is a simple, humble enhancement.' According to company legend, the brand owes its name to the chemist Norm Larsen. He was trying to invent a liquid that would prevent corrosion, hence 'Water Displacement' in the name. WD-40 was his fortieth attempt. It is impossible, of course, to know how well the product would have done had Larsen got the formula right on his forty-first try. Yet academic research corroborates Greg's semiotic evaluation: for household products, divisible numbers are more attractive to consumers than indivisible ones.

In 2011, Dan King of the National University of Singapore and Chris Janiszewski of the University of Florida demonstrated that an imaginary brand of anti-dandruff shampoo was better liked when it was called Zinc 24 than when it was called Zinc 31. The respondents preferred Zinc 24 so much that they were willing to pay ten per cent more for it. King and Janiszewski argued that customers prefer 24 because they are more familiar with the number from their schooldays, when the lines $3 \times 8 = 24$ and $4 \times 6 = 24$ are drummed into pupils by rote. By comparison, 31 is a prime number and does not appear in any school multiplication table. The professors claimed that increased familiarity with 24 means we process the number more fluently, which gives us the feeling that we like it more. Our preference for 24 over 31, they argued, transfers to a preference for Zinc 24 over Zinc 31. Greg was not surprised when I told him of this research, but he had a more cultural take: 'Zinc 24 fits our sense that even-numbered products bring us back to a sense of normalcy, to a sense of things as they should be,' he said. 'Odd numbers provide extra room for a bit of emotional negotiation, which is why there is more mysticism around them.' And why, he added, we don't want them in our hair.

To reinforce their hypothesis that processing fluency increases brand preference, King and Janiszewski designed a follow-up experiment that subtly included a multiplication sum in the advertisement for a numbered brand. They first decided on the products, Solus 36 and Solus 37, two fictitious lines of the real contact lens brand Solus. They then created four ads: one for Solus 36, one for Solus 37, and one for each product with the tag line '6 colors. 6 fits.' When there was no tag line, the participants preferred Solus 36 over Solus 37, as would be expected. But when the researchers included the tag line, Solus 36 increased in popularity and Solus 37 became even less popular than before. King and Janiszewski argued that our familiarity with 6, 6 and 36, from the six times table sum $6 \times 6 = 36$, increases our processing fluency of the numbers, just as the unfamiliarity of 6, 6 and 37, which are not arithmetically related, decreases it. The pleasure rush that comes from subconsciously recognizing a simple multiplication makes us feel good, they said, and we misattribute the buzz as satisfaction with the product. Companies would do well, they suggested, to include hidden sums in their ads.

Which packet of contact lenses feels more desirable?

King and Janiszewski's point is that we are always sensitive to whether a number is divisible or not, and this sensitivity influences our behaviour. We are all a bit like Jerry Newport, the taxi driver from Tucson, who cannot see a number without dividing it up into primes. Splitting in two is the earliest and most natural type of division. The arithmetical pattern, therefore, that we are most sensitive to – and for which our cultural associations are the most deeply rooted – is the difference between the evens and the odds.

Numbers were invented to describe precise amounts: three teeth, seven days, twelve goats. When quantities are large, however, we do not use numbers in a precise way. We approximate using a 'round number' as a place mark. It is easier and more convenient. When I say, for example, that there were a hundred people at the market, I don't mean that there were *exactly* one hundred people there. And when I say that the universe is 13.7 billion years old, I don't mean *exactly* 13,700,000,000, I mean give or take a few hundred million years. Big numbers are understood approximately, small ones precisely, and these two systems interact uneasily. It is clearly nonsensical to say that next year the universe will be '13.7 billion and one' years old. It will remain 13.7 billion years old for the rest of our lives.

Round numbers usually end in zero. The word *round* is used because a round number represents the completion of a full counting cycle, not because zero is a circle. There are ten digits in our number system, so any combination of cycles will always be divisible by ten.

Because we are so used to using round numbers for big numbers, when we encounter a big number that is non-round – say, 754,156,293 – it feels discrepant. Manoj Thomas, a psychologist at Cornell University, argues that our sense of unease with large, non-round numbers causes us to see these numbers as smaller than they are: 'We tend to think that small numbers are more precise, so when we see a big number that is precise we instinctively assume it is less than it is.' The result, he claims, is that we will pay more for an expensive object if the price is non-round. In one of Thomas's experiments, respondents viewed pictures of several houses together with their sale prices, which were randomly assigned either a round number, such as $390,000, or a slightly larger, precise one, such as

$391,534. Asked whether they considered each price high or low, on average the respondents judged the precise prices to be lower than the round ones, even though the precise numbers were actually higher. Thomas and his collaborators argued that whatever other inferences the respondents were making about why the price was precise – such as that the seller had thought more carefully about it, and so the price was fairer – they still made the subconscious judgement that non-round numbers are smaller than round ones. A tip to readers selling their homes: if you want to make money, don't end the price with a zero.

Earlier we discussed the cultural connotations of adding 1 to a round number. The practice of *subtracting* 1 from a round number also conveys a potent message.

When we read a number, we are more influenced by the leftmost digit than we are by the rightmost, since that is the order we read, and process, them. The number 799 feels significantly less than 800 because we see the former as 7-something and the latter as 8-something, whereas 798 feels pretty much like 799. Since the nineteenth century, shopkeepers have taken advantage of this trick by choosing prices ending in a 9, to give the impression that a product is cheaper than it is. Surveys show that anything between a third and two-thirds of all retail prices now end in a 9.

Though we are all seasoned shoppers, we are still fooled. In 2008, researchers at the University of Southern Brittany monitored a local pizza restaurant that was serving five types of pizza at €8 each. When one of the pizzas was reduced in price to €7.99, its share of sales rose from a third of the total to a half. Dropping the price by one cent, an insignificant amount in monetary terms, was enough to influence customers' decisions dramatically.

Our response to prices ending in 9, however, is subject to more complex influences than a bias towards the leftmost digit. A price ending in 9 feels like a bargain, even when it isn't. Eric Anderson of the University of Chicago and Duncan Simester of MIT arranged for the same dress to be priced at $34, $39 and $44 in three otherwise identical mail order catalogues. The dress sold best at $39, rather than the cheaper price of $34. Similar results have been found in other studies: the 9-ending is a cue that the item has been discounted and is therefore a good deal. But the association of 9 with bargains can also mean that a 9-priced product looks cheap, or can give the impression that the seller is somehow manipulating you. An upmarket restaurant, for example, would never dream of pricing a main course at, say, £22.99. Nor would you trust a therapist who charged £59.99 a session. The prices would be £23 and £60, which feel both classier and more honest. Our response to the number 9 is conditioned by a mixture of cultural and psychological factors. Numbers are not impartial and straightforward; they have baggage.

Shopkeepers have other reasons for using prices that end in a 9, or, for that matter, an 8. Tests show that prices ending in 8 and 9 are much harder to recall than prices ending in 0 and 5, since the brain takes longer to store and process them. If you don't want your customers to remember a price, to prevent them comparing prices, then use an 8- or 9-ending. Conversely, if you *do* want a customer to remember a price, perhaps to reinforce that it is cheaper than rival products, label it £5 and not £4.98. Traders, in fact, use an array of psychological number tricks to reduce price awareness. For example, a Cornell University study showed that by leaving the currency sign off a menu – so that the price of a dish was listed as 20 rather than $20 – a New York restaurant increased average spend per customer by eight per cent. The '$' reminds us of the pain of paying. Another clever menu strategy is to show the prices immediately after the description of each dish, rather than listing them in a column, since listing prices facilitates price comparison. You want to encourage diners to order what they want, whatever the price, rather than reminding them which dish is most expensive.

Perhaps the most blatant use of number psychology in retail, however, is the display of absurdly expensive items to create an artificial benchmark. The £100,000 car in the showroom and the £10,000 pair of shoes in the shop window are there not because the manager thinks they will sell, but as decoys to make the also-expensive £50,000 car and £5,000 shoes look cheap. Supermarkets use similar strategies. We are surprisingly susceptible to number cues when it comes to making decisions, and not just when shopping. In one study, 52 German judges read a description of

Roast fillet of smoked haddock with warm potato salad
and crispy onions 7.50
Cream of mushroom soup with truffle chantilly 5.50
Warm organic Gloucestershire chicken ballotine stuffed with
herbed couscous and leek fondue 8.20

Roast fillet of smoked haddock with warm potato
salad and crispy onions £7.50

Cream of mushroom soup with truffle chantilly £5.50

Warm organic Gloucestershire chicken ballotine
stuffed with herbed couscous and leek fondue £8.20

Good menu, bad menu.

a woman caught shoplifting, and then rolled a pair of dice, loaded
to land either on 1 and 2, or on 3 and 6. Once the dice were rolled,
the judges were asked to state whether they would sentence the
girl to more or fewer months in prison than the sum of the numbers
on the dice, and then to specify the exact sentence. The judges
who rolled 3 gave her five months on average, while the judges who
rolled 9 gave her eight. The judges were experienced professionals,
yet the mere suggestion of a number with no connection to the
case determined the length of sentence.

If earnest German judges can be swayed by an irrelevant random
number, think about the rest of us. Every time we perceive a number
it primes us, influencing our behaviour in ways we are not always
aware of and cannot always control.

Another response to numbers is affection. After counting, calculating
and quantifying with our numerical tools it is common to develop
feelings for them. Jerry Newport, for example, loves some numbers
like friends. I had not realized the depth of our collective number
love, however, until I conducted an online experiment, asking
members of the public to nominate their favourite numbers and
explain their choices. I was taken aback not only by the level of

interest – more than 30,000 people took part in the first few weeks – but also by the variety and tenderness of the submissions: 2, because the respondent has two piercings; 6, because the sixth track on the respondent's favourite albums is always the best song; 7.07, since the respondent used to always wake up at 7.07am, and once her shopping added up to $7.07 in front of the cute cashier at her local shop; 17, because that's how many minutes the respondent takes to cook rice; 24, because the respondent sleeps with her left leg kicked out like a 4 and her boyfriend sleeps like a 2 on his side; 73, known to fans of *The Big Bang Theory* as the 'Chuck Norris of numbers', because the main character, Sheldon Cooper, points out that it is the 21st prime number, and its mirror 37 is the 12th; 83, because it sounds good to exaggerate with, as in 'I must have done it 83 times!'; 101, because it is the lowest whole number with an 'a' in it; 120, because it is divisible by 2, 3, 4, 5, 6, 8 and 10, providing the respondent with sufficient numbers to count up and down to get to sleep; 159, because it is the diagonal on a phone keyboard; 18,912, because its cadence makes it 'the most beautiful sounding number in the world'; and 142,857, the phoenix number, because its first six multiples are well-ordered numerical anagrams of itself:

$$
\begin{aligned}
&\qquad\quad\ 142857142857 \\
142857 \times 1 &= 142857 \\
142857 \times 2 &= 285714 \\
142857 \times 3 &= 428571 \\
142857 \times 4 &= 571428 \\
142857 \times 5 &= 714285 \\
142857 \times 6 &= 857142 \\
142857 \times 7 &= 999999
\end{aligned}
$$

'Having a favourite number means that you get a little buzz every time you happen to be sitting in seat 53 on a train, or notice that the time is 09:53,' wrote one respondent. 'I can't think of a reason not to have a favourite number.'

With the caveat that the survey was voluntary and self-selecting, a bit of fun rather than rigorously undertaken academic research, the data revealed fascinating patterns in favourite number choices.

Firstly, the span of our number hug is huge: 1123 individual numbers from 30,025 submissions. There were votes for every whole number between 1 and 100, and 472 of the numbers between 1 and 1000. The lowest whole number that failed to pick up any votes was 110. Surely the world's least-loved number?

Here's the final table:

Position	Number	Percentage
1	7	9.7%
2	3	7.5%
3	8	6.7%
4	4	5.6%
5	5	5.1%
6	13	5.0%
7	9	4.8%
8	6	3.4%
9	2	3.4%
10	11	2.9%
11	42	2.8%
12	17	2.7%
13	23	2.3%
14	12	2.2%
15	27	1.9%
16	22	1.5%
17	21	1.4%
18	π	1.4%
19	14	1.3%
20	24	1.2%
21	1	1.2%
22	16	1.2%
23	10	1.2%
24	37	1.0%
25	0	1.0%
26	19	0.9%
27	18	0.8%
28	e	0.7%
29	28	0.7%
30	69	0.6%

Roughly speaking, we like single digits best, and the bigger a number is, the less we like it. The table also reveals a shocking indifference towards round numbers. The numbers from two to nine are all in the top ten, but ten is way down in 23rd place, twenty is in 50th and thirty in 69th. Ten is the cornerstone of the decimal system, yet it is not very lovable, possibly because it is always prostituting itself as an approximation.

Some numbers are chosen for their numerical properties, such as the phoenix number on p. 14, and also the constants π and e, which we will look at much more closely later in this book. Usually, however, a number is chosen for a personal reason, most commonly because it is the day of the month we were born. Yet the distinction between a numerical and a personal reason is not clear-cut, since there are some numbers that are rarely chosen as favourites even if the person was born on that day. For example, if you were born on the 10th of the month, you are six times less likely to choose 10 as your favourite number than you are likely to choose 7 if you were born on the 7th of the month. If you were born on the 30th you are forty times less likely to choose 30. Some numbers evidently make better favourites than others. (One of the reasons I became so curious about favourite numbers is because I don't have one, and I couldn't quite believe that so many other people felt so passionately about them. Now I blame my lack of a favourite number on the fact that I was not born between the 2nd and the 9th of the month.)

The historic tendency of odd numbers to attract more attention than even numbers is reflected in the survey. Among the submissions for numbers between 1 and 1000, the ratio of those preferring odds over evens is about 60:40. The table also shows that Douglas Adams' joke that 42 is the answer to life, the universe and everything is still hilarious more than three decades after he first made it. (His gag plays on our collective feelings about numbers too: 42 works because it is so bland. It would not be as funny had he chosen, say, 41, which is odd and prime.) The appearance of 69 shows that juvenile humour cannot be eliminated from internet polls.

———

Seven came first overall. It was also the unanimous choice irrespective of the age, gender and mathematical ability of the respondent, which is hardly a surprise. Seven has been the most culturally feted number for as long as we know. Wonders of the world, deadly sins, ages of man, pillars of wisdom, brides for brothers, seas, samurai and dwarves all come in sevens. Babylonian ziggurats were built with seven storeys, the Egyptians spoke of the seven gates of the netherworld, the Vedic sun god has seven horses, and Muslims must walk round the Kaaba seven times during the Hajj. Even now, the fundamental rhythm of our lives is a cycle of seven: the number of days in a week.

The very first thing humans counted was time. We carved notches on sticks and daubed splotches on rocks to mark the passing of days. Our first calendars were tied to astronomical phenomena, such as the new moon, which meant that the number of days in each calendar cycle varied, in the case of the new moon between 29 or 30 days, since the exact length of a lunar cycle is 29.53 days. In the middle of the first millennium BCE, however, the Jews introduced a new system. They decreed that the Sabbath come every seven days ad infinitum, irrespective of planetary positions. The continuous seven-day cycle was a significant step forward for humanity. It emancipated us from consistent compliance with Nature, placing numerical regularity at the heart of religious practice and social organization, and since then the seven-day week has become the world's longest-running uninterrupted calendrical tradition.

But why *seven* days in the week? Seven was already the most mystical of numbers by the time the Jews declared that God took six days to make the world, and rested the day after. Earlier peoples had also used seven-day periods in their calendars, although never repeated in an endless loop. The most commonly accepted explanation for the predominance of seven in religious contexts is that the ancients observed seven planets in the sky: the Sun, the Moon, Venus, Mercury, Mars, Jupiter and Saturn. Indeed, the names Saturday, Sunday and Monday come from the planets, although the association of planets with days dates from Hellenic times, centuries after the seven-day week had been introduced. It is ironic that the Jewish week – the first calendar system to sever the link between the planetary orbits and the counting of days – ended up with its seven days named after the planets. Perhaps the astrological connection made the week more resilient to competing systems. Some historians argue that a period of seven days was originally chosen because it is roughly a quarter of a 29.53-day lunar month. But if divisibility were the issue a more accurate calendar would have had five weeks of six days, six weeks of five days, or even three weeks of ten days.

The Egyptians used the following hieroglyph for seven, 𓁶, the human head, which suggests another possible reason for the number's symbolic importance. There are seven orifices in the head: the ears, eyes, nostrils and mouth. Human physiology provides other explanations too. Six days might be the optimal length of time to work before you need a day's rest, or seven might be the most appropriate number for our working memory: the number of things the average person can hold in his or her head simultaneously is seven, plus or minus two.

I'm not convinced by any of the reasons above, even if they are happy coincidences. Seven is special not because of planets, orbits or orifices, but because of arithmetic. Seven is unique among the first ten numbers because it is the only number that cannot be multiplied or divided within the group. When 1, 2, 3, 4 and 5 are doubled the answer is less than or equal to ten. The numbers 6, 8 and 10 can be halved and 9 is divisible by three. Of the numbers we can count on our fingers, only 7 stands alone: it neither produces nor is produced. Of course the number feels special. It is!

Psychologists have studied the uniqueness of seven for decades. When people are asked to think of a digit off the top of their heads, they are most likely to think of a 7. When asked to think of a number between 1 and 20, the majority will think of 17. Such is the subconscious drive towards numbers ending in 7 that it is the basis of a classic trick, in which the magician asks a volunteer to think of a two-digit odd number between 1 and 50 whose digits are different (so 15 is permitted but 11 is not), and correctly predicts that he or she is thinking of … the number in the footnote overleaf. Have a guess before you look.* The psychologists

Michael Kubovy and Joseph Psotka argued that when asked to generate a random digit, participants will eliminate numbers that seem too unspontaneous – the even numbers, the multiples of three, and the numbers 0, 1 and 5 since they fall either at the beginning or the middle of the sequence. Seven is the oddest man out – non-even, non-round and prime.

A favourite number reflects one's uniqueness. You can't do better than seven, the ultimate outsider.

Numbers express *quantities*. In the submissions to my online survey, however, respondents frequently attributed *qualities* to them. Noticeably, colours. The number that was most commonly described as having its own colour was four (52 votes), which most respondents (17) said was blue. Seven was next (28 votes), which most respondents (9) said was green, and in third place came five (27 votes), which most respondents (9) said was red. Seeing colours in numbers is a manifestation of synaesthesia, a condition in which certain concepts can trigger incongruous responses, and which is thought to be the result of atypical connections being made between parts of the brain.

In the survey, numbers were also labelled 'warm', 'crisp', 'chagrined', 'peaceful', 'overconfident', 'juicy', 'quiet' and 'raw'. Taken individually, the descriptions are absurd, yet together they paint a surprisingly coherent picture of number personalities. Below is a list of the numbers from one to thirteen, together with words used to describe them taken from the survey responses.

One	Independent, strong, honest, brave, straightforward, pioneering, lonely.
Two	Cautious, wise, pretty, fragile, open, sympathetic, quiet, clean, flexible.
Three	Dynamic, warm, friendly, extrovert, opulent, soft, relaxed, pretentious.
Four	Laid-back, rogue, solid, reliable, versatile, down-to-earth, personable.

Five	Balanced, central, cute, fat, dominant but not too much so, happy.
Six	Upbeat, sexy, supple, soft, strong, brave, genuine, courageous, humble.
Seven	Magical, unalterable, intelligent, awkward, overconfident, masculine.
Eight	Soft, feminine, kind, sensible, fat, solid, sensual, huggable, capable.
Nine	Quiet, unobtrusive, deadly, genderless, professional, soft, forgiving.
Ten	Practical, logical, tidy, reassuring, honest, sturdy, innocent, sober.
Eleven	Duplicitous, onomatopoeic, noble, wise, homey, bold, sturdy, sleek.
Twelve	Malleable, heroic, imperial, oaken, easy-going, non-confrontational.
Thirteen	Gawky, transitional, creative, honest, enigmatic, unliked, dark horse.

You don't need to be a Hollywood screenwriter to spot that Mr One would make a great romantic hero, and Miss Two a classic leading lady. The list is nonsensical, yet it makes sense. The association of one with male characteristics, and two with female ones, also remains deeply ingrained.

Since the favourite-number survey was voluntary, it was biased towards those people who already had clear emotional attachments to numbers. But what about everyone else?

Take the number 44.

Do you like it? Do you dislike it? Do you remain unmoved?

Dan King and Chris Janiszewski, the professors we met earlier in our discussion of the shampoo Zinc 24, conducted an experiment

---

* Most people think of 37.

in which all participants indicated whether they liked, disliked or felt neutral about every number from 1 to 100. The numbers were ranked from most liked to least liked.

The responses demonstrated that this was not a ridiculous question to ask. Our liking of numbers follows clear patterns, as shown below in a 'heat map' in which the numbers from 1 to 100 are represented by squares. (The top row of each grid contains the numbers 1 to 10, the second row the numbers 11 to 20, and so on.) The numbers marked with black squares represent those that are 'most liked' (the top twenty in the rankings), the white squares are the 'least liked' (the bottom twenty) and the squares in shades of grey are the numbers ranked in between.

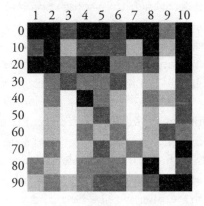

The heat map shows conspicuous patches of order. Black squares are mostly positioned at the top of the grid, showing on average that low numbers are liked best. The left sloping diagonal through the centre reveals that two-digit numbers where both digits are the same are also attractive. We like patterns. Most strikingly, however, four white columns display the unpopularity of numbers ending in 1, 3, 7 and 9. King and Janiszewski's opinion, as mentioned previously, is that numbers that are the answers to common arithmetical problems, such as numbers that appear in the times tables, are more familiar, more fluently processed and hence more liked. The even numbers and numbers ending in 5 are *always* divisible, but the numbers ending in 1, 3, 7 and 9 are often not.

In a similar experiment, Marisca Milikowski of the University of Amsterdam asked participants to rank each number between

1 and 100 on three scales: between good and bad, between heavy and light, and between excitable and calm. Again, when asked to project non-mathematical meanings onto numbers, our responses are remarkably coherent. I have translated the results into the heat maps opposite.

The patterns are pronounced. The white columns in the 'Good' grid show that numbers ending in 3, 7 and 9 are the least good, which is perhaps not surprising, since we saw previously that we like them least. In the 'Heavy' grid, the black has sunk to the bottom, indicating that the larger a number is, the heavier we interpret it to be. The pattern in the 'Excitable' grid is not obvious at first, but upon examination the columns ending in an odd number are darker on average than those ending in an even number. Odds are excitable, and evens calm. We find it easy to project non-mathematical meanings onto numbers, and these meanings reflect numerical properties, most clearly size and divisibility.

The bottom left grid is a heat map of number rankings from the favourite number survey, with the top 20 favourites in black, and so on. The bottom right grid displays the results of another online survey I set up, in which participants chose a number at random between 1 and 100. The 20 most popular nominations are in black. Interestingly, these two maps resemble each other: when asked to think about which number we like best, and when asked to think of the first number that comes into our heads, we tend to nominate the same candidates. Counter-intuitively, our favourite numbers are generally not the numbers that we like best or think are most good. Like is very different from *love*.

When I first saw the heat maps, I instantly thought of Jerry Newport, the world champion mental calculator and former taxi driver I visited in Arizona. Jerry told me that when he comes face to face with a four or five digit number, he spontaneously 'sieves' out the prime numbers. In other words, he initially calculates if it is possible to divide the number by 2, and then by 3, and then by 5, 7, 11, and so on upwards, in order to decompose it into its unique prime divisors.

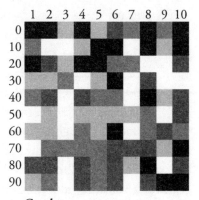

Good

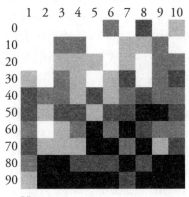

Heavy

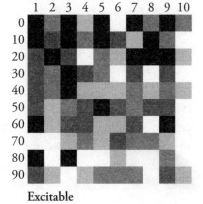

Excitable

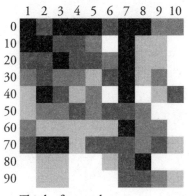

Favourite

Think of a number

As we saw:

$$2761 = 11 \times 251$$
$$2762 = 2 \times 1381$$
$$2763 = 3 \times 3 \times 307$$

The heat maps brought it home to me that we are all sieving prime numbers. The facing page has the original grids with the primes marked on them. They look like sieves! In 'Liked' and 'Good', the primes fit almost perfectly into the white spaces, as if they are falling through holes in a wire gauze. Conversely, the majority of the primes in 'Excitable', 'Favourite' and 'Think of a number' fall on the black and dark grey squares, as if these grids are sieves designed to catch them. The prime numbers are significant features of our internal landscape of numbers, not just for savants like Jerry Newport, but for the rest of us, too. Our brains are always switched on to arithmetic.

Numbers assault us at every moment of the day. They holler from clocks, phones, newspaper pages, computer screens, street signs, price tags, bus stops, addresses, licence plates, billboards and books. As we've seen in this chapter, they ceaselessly needle our neurons. And when we look right back at them, amazing patterns come into view.

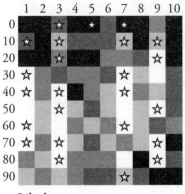

Liked

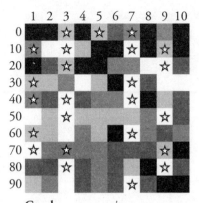

Good

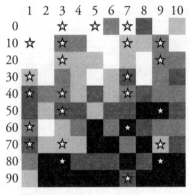

Heavy

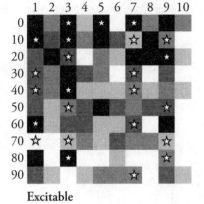

Excitable

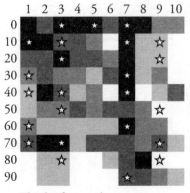

Favourite

Think of a number

Heat maps with the prime numbers starred.

The Long Tail of the Law

In 1085 William the Conqueror ordered a survey of England. He wanted to know how many people lived in his land, who they were, what they owned, how much they were worth and – most importantly – what taxes they paid. He dispatched officials to all corners of his kingdom, where his demands were carried out so thoroughly that the *Anglo-Saxon Chronicle* declared, 'not one ox nor one cow nor one pig was left unrecorded.'

The results became known as the *Domesday Book*. It is Britain's earliest compendium of population statistics, the first large data set in the Western world, and a trove for historians, geographers, genealogists and lexicographers. Curious to see if it concealed any mathematical secrets, I examined the first section, which covers the county of Kent.

The opening line reports that the town of Dover rendered £18, of which 2 parts had gone to King Edward and the third to Earl Goodwine. The inhabitants of Dover gave the King 20 ships for 15 days, with 21 men each.

Since I was concerned only with the numbers, I read the above passage as the list 18, 2, 20, 15 and 21. I noticed something straight away. Look at the *first digit* of each number. Taken together they are 1, 2, 2, 1 and 2. Only 1s and 2s. The lowest digits. Interesting? Possibly, yes. Still, it was a small sample, too small to make any conclusions. I carried on through the book, tallying the first digits of all the numbers that came up. The abundance of 1s and 2s continued. Yes, 3s and 4s and the others revealed their squiggly faces, but more sheepishly. It was striking to observe how much more frequently numbers started with low digits than with high ones.

I had to count 182 numbers before arriving at the first to begin with a 9. It was a single 9, the number of a type of peasant bound to Wulfstan son of Wulfwine of Shepherdswell. By that time I had counted 53 numbers beginning with a 1, 22 beginning with a 2, 18 beginning with a 3 and 15 beginning with a 4. Look at these numbers again. A clear pattern has emerged. Numbers beginning with 1 are more common than numbers beginning with 2, which are more common than numbers beginning with 3, and so on down to the least common first digit, 9.

I could see why 1 was so popular. The Domesday inspectors moved from dwelling to dwelling, enumerating humans, livestock and possessions. Farms that had ploughs usually had just one of them. Inevitably the 1 was well represented. Yet this did not explain the remarkably consistent drop in frequencies as the first digit increased, especially when the numbers were referring to such different stuff in such different quantities, such as the 40,000 herrings given to monks at Canterbury or the 27 salt pans in Milton Regis.

Perhaps it was a medieval thing. I closed the Domesday Book and fast-forwarded my research 800 years.

I alighted in Victorian London. On 12 March 1881, the front page of *The Times* announced that the owner of a 25-ton schooner was requesting to meet with gentlemen to accompany him on a tour of the South Seas; the Temporary Home for Lost Dogs in Battersea was asking if persons desirous of purchasing pets might view its 500 to 700 canine residents; and Samuel Brandram begged to announce his Tuesday afternoon Shakespearean recitals at 3pm, at 33 Old Bond-street, with reserved seating at 5 shillings.

I counted the frequencies of the first digits – also called the *leading digits* – of all the numbers on the front page. Again numbers beginning with a 1 were the most common and numbers beginning with a 9 were the least common. Life in the nineteenth century was different from that in the eleventh, yet the leading digits of its social statistics behaved almost identically.

If you take any newspaper today, you will find the same phenomenon. Try it! It's a simple party trick and a pub magician's stock-in-trade. Count the leading digits and you will discover that they always follow a sliding scale, with 1 far and away the most abundant, 2 the next in line, 3 after that, and so on all the way down to 9, the rarest of the pack.

It's startling. Most people won't believe you until you tally the digits. Instinctively, we do not expect newspaper numbers to be so well-behaved, especially since they are determined so randomly and from such wide-ranging sources. Whether the numbers come from sports scores, share prices or death tolls, I guarantee you the most common leading digit will be 1 and the least common will be 9.

The result is surprising because we would intuitively expect every number to have an equal chance of occurring. Indeed, when numbers are selected randomly from a box of 999 bouncing ping pong balls displaying the numbers from 1 to 999, the chances of selecting one whose number begins with any particular digit is exactly a ninth, or 11 per cent. Each digit has just as much chance of being chosen as any other. Yet, demonstrably, leading digits in newspapers are not distributed this way. The first digits dance to a distinctive and asymmetric beat.

The curious profusion of numbers beginning with a 1 was first noticed by the Canadian-born American astronomer Simon Newcomb. In 1881 he published a brief note in the *American Journal of Mathematics,* saying that he had been led to his conclusion by observing fraying books of logarithm tables. The first pages, with tables of numbers beginning with a 1, were always more worn out than the last ones, containing tables beginning with a 9. The phenomenon wasn't because researchers were reading the log tables from the beginning and then giving up a few pages in for lack of a gripping narrative. Numbers beginning with a 1 appeared most often in their work. Newcomb conjectured that the frequencies of first digits follow these percentages:

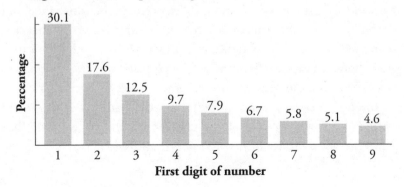

The number 1 appears 30.1 per cent of the time, 2 appears 17.6 per cent of the time and 3 appears 12.5 per cent of the time. The drop in frequencies is so dramatic that 1s are almost seven times more likely to occur than 9s.

Newcomb's percentages are derived from logarithms. The probability of a number beginning with the digit d, he argued, is $\log (d +1) - \log d$. (I explain what this means in Appendix One on p. 291). He did not prove rigorously why this was so but instead used a short, informal argument, which he presented as a curiosity.

More than half a century later, in 1938, the physical condition of books of log tables led Frank Benford, a physicist at the General Electric Company in New York – apparently unaware of Newcomb's paper – to 'rediscover' the first digit phenomenon. Benford, however, investigated more than just books of logs, and discovered a surfeit of 1s and a paucity of 9s wherever he looked. He tallied the first digits from US city population tables, from the addresses of the first few hundred people in *American Men of Science*, from the atomic weights of elements, from tables of the areas of rivers and from baseball statistics. Most of the data sets were fairly close to the expected distribution. It must have been mind-blowing to see the same percentages emerge from such different situations. Of course, the percentages were not the exact ones above (the real world is not precise like that), but when taken together the data tallied with the predicted values with only a small margin of error, just a few tenths of one per cent. Now it is an accepted part of physical science, finance, economics and computing that in data culled from naturally occurring, random processes spread over several degrees of magnitude, the leading digit will be a 1 about 30 per cent of the time, a 2 about 18 per cent of the time, and so on. Benford argued that the phenomenon must be evidence of a universal law, which he called the Law of Anomalous Numbers. The coinage didn't catch on. His name, however, did. The phenomenon is known as Benford's law.

The law is one of many notable number patterns in the world around us. I will arrive at some others in a few pages' time, but before we get there we have some sleuthing to do.

———

Many real-world data sets closely follow Benford's law, such as the populations of all 3221 US counties, and the total quarterly revenues of 30,525 US public firms between 1961 and 2011.

Darrell D. Dorrell reminded me of a bear. This association was partly because we met in Portland, capital of the bear-rich state of Oregon, and partly because of his stocky physique, stubbly moustache and softly rumbling voice. It was also because of the nature of his job as a financial investigator. Darrell sniffs out corrupt data with the predatory instincts of a grizzly looking for dinner. You don't want him going through your books if there is even the slightest scent of wrongdoing. The CIA, the Department of Justice and the Securities and Exchange Commission have all engaged Darrell's services for 'forensic accounting', the industry term for investigating financial malpractice. He has a licence to carry a gun at all times. 'All our doors in here lock internally,' he said. 'We make a lot of people unhappy.'

When Darrell first heard of Benford's law in the early noughties, he went through the same process one gets after experiencing major loss: surprise, denial, anger and then acceptance. 'I first thought "why have I never heard about this before?" Then I thought "It can't be true!" And once I had gotten to understanding it, I had the epiphany of "Gee! This is another tool we can use."' Now one of the first things that Darrell does when he investigates fraud is to run a check on the leading digits of bank accounts and company ledgers. Financial data that is spread over several orders of magnitude – that is, in which there are entries for amounts measured in units, tens, hundreds and thousands of dollars – will conform to Benford's law. If it doesn't there could be a legitimate explanation, such as recurring purchases of an item worth, say, $40, which would give a blip on 4. Or there could be malpractice. Divergence from the Benford's percentages is a red flag that demands a closer look.

Darrell pointed to a framed newspaper front page on his wall, announcing the conviction of Wesley Rhodes, a local financial advisor who stole millions of dollars from investors to buy classic cars. 'Benford's helped us prosecute him,' said Darrell. The statements that Rhodes sent investors failed the first-digit test, indicating that something was not right. Under closer examination, Darrell discovered that Rhodes had faked the data. Darrell now calls Benford's law 'the DNA of quantitative investigation. It's the basic premise of how our digits work. And – as I have explained in court several times – the nice thing is that it is science. Benford's is not a theory. It's a *law*.'

The method of screening digits for compliance with Benford's law is increasingly used to detect data manipulation, not only in cases of financial fraud but also wherever it is reasonable to assume that the law applies. Scott de Marchi and James T. Hamilton of Duke University, North Carolina, wrote in 2006 that self-reported data from industrial plants about levels of lead and nitric acid emissions did not follow Benford's law, suggesting that the plants were not providing accurate estimates. Walter Mebane, a political scientist at the University of Michigan, used Benford's law to claim that the Iranian presidential election in 2009 was rigged. Mebane analysed the ballot-by-ballot results and found that the votes for conservative incumbent Mahmoud Ahmadinejad showed high discrepancy with the law, whereas the votes for his reformist challenger Mir Hossein Mousavi did not. 'The simplest interpretation,' he wrote, 'is that votes were somehow artificially added to Ahmadinejad's totals while Mousavi's counts remained unmolested.' Scientists also use Benford's law as a diagnostic tool. During an earthquake, the measurements of the peaks and troughs of a seismogram follow the law. Malcolm Sambridge, of the Australian National University, analysed two different seismograms that recorded the 2004 Indonesian earthquake – one in Peru and one in Australia. The former followed Benford's law but the latter did not adhere as closely. From this he deduced that a minor seismic disturbance near Canberra must have interfered with the data. The first-digit test revealed an earthquake that had gone unnoticed.

Not only is 1 more common than 2 as a leading digit, but 1 is also more common than 2 in the second, third, fourth and indeed *any* position in a number. The illustration below has the Benford's percentages for second digits (which now include 0). The differences are not as pronounced as for leading digits, but they are sufficient for many diagnostic uses, such as analysing financial and electoral data. The further along a number you go the more similar the percentages become. Benford's is not just a first digit phenomenon. There really *are* more 1s in the world!

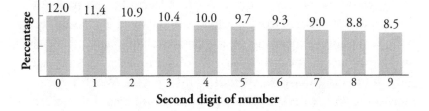

In court, Darrell D. Dorrell is often called to explain why Benford's law is true. He stands up in front of a flipchart and starts counting upwards from 1, writing down the digits and saying them aloud. He feels like a schoolteacher giving a maths class. 'The judge, the opposing attorney – it drives them crazy!' he said.

We can do a similar exercise. Here are the numbers from 1 to 20:

1, 2, 3, 4, 5, 6, 7, 8, 9, 10, 11, 12, 13, 14, 15, 16, 17, 18, 19, 20

More than half the digits start with a 1, since between 11 and 19 they all do. Continue counting. Wherever we choose to stop we will always have passed at least as many numbers beginning with a 1 as start with a 2, since to get to the twenties, or two hundreds, or two thousands, we will necessarily have counted through the teens, hundreds and thousands. Likewise there are at least as many numbers beginning with a 2 as start with a 3, and so on all the way up to numbers beginning with a 9. This argument, however, while it gives us an intuitive feel for why Benford's is true, and is sufficient for a court of law, is not a proof in the stricter court of mathematics.

———

One of the most striking aspects of Benford's law is that the pattern is independent of the units of measurement. If a set of financial data follows Benford's law when measured in pounds, it will also follow the law when all amounts are converted into dollars. If a set of geographical data conforms to Benford's in kilometres, it will also conform when the data is presented in miles. This property – called *scale invariance* – is necessarily true, since numbers taken from newspapers, bank accounts and atlases all over the world show the same distribution of first digits, irrespective of the currency and systems of measure being used.

To change a number from miles to kilometres you multiply it by 1.6, and to change from pounds to dollars you also multiply by a fixed number, depending on the day's exchange rate. The easiest way to appreciate the scale invariance of Benford's law is to consider how numbers behave when multiplied by two. When a number beginning with a 1 is multiplied by 2, the result begins with a 2 or a 3. (For example, $12 \times 2 = 24$ and $166 \times 2 = 332$.) When a number beginning with a 2 is multiplied by 2, the result begins with a 4 or a 5. (For example, $2.1 \times 2 = 4.2$ and $25 \times 2 = 50$.) The first two lines of the table below show what happens to the leading digit of a number when the number is doubled:

First digit of n	1	2	3	4	5	6	7	8	9
First digit of $2n$	2 or 3	4 or 5	6 or 7	8 or 9	1	1	1	1	1
Benford's percentage	30.1	17.6	12.5	9.7	7.9	6.7	5.8	5.1	4.6

Now let S be a set of numbers that obeys Benford's law, and let's multiply every number in S by 2, and call our new set T. The table tells us that the numbers in S beginning with a 5 make up 7.9 per cent of the set, that the numbers beginning with a 6 make up 6.7 per cent, and that the numbers beginning with a 7, 8 or 9 make up 5.8, 5.1 and 4.6 per cent respectively. So, the percentage of numbers in S that begin with a 5, 6, 7, 8 or 9 is 7.9 + 6.7 + 5.8 + 5.1 + 4.6

= 30.1 per cent. When all the numbers beginning 5, 6, 7, 8 or 9 are doubled, the result always begins with a 1, as the table shows. In other words, 30.1 per cent of the numbers in T begin with a 1, which is the Benford's percentage!

The figures also tally when we consider the other digits. Multiplying by 2 dilutes and re-concentrates the Benford's blend perfectly, preserving the distribution of first digits. I chose to multiply by 2 since it is the easiest possible multiplier, but I could equally have multiplied by three or by 1.6 or by pi, or by any other number, and the set's Benfordness would have remained. Under any change of scale the distribution always reconfigures itself, as if stirred by a Divine hand.

For decades after Frank Benford discovered it, his law was considered a quirk of data, a gimmick for magic shows, numerology rather than mathematics. In the nineties, however, Ted Hill, a professor at Georgia Tech, set his sights on finding a theoretical explanation for its prevalence. He now lives in Los Osos, California, down the Pacific coast from Darrell D. Dorrell. Ted is a former soldier who retains a military air: tall, broad-shouldered and lean, with a shaved head and a white moustache. When I visited him, he led me to a small wooden cabin at the end of his garden, overlooking the ocean and two national parks. A log fire was crackling. He calls the cabin his 'math dacha'. It is the global epicentre of mathematical research into Benford's law.

Ted's first major result was to prove that if there is a universally observed pattern for leading digits, then Benford's law is the only possible candidate. He did this by showing that the only pattern of first digits that stays the same when the scale is changed is Benford's. His insight led him to invent the following game, which he played with me:

'You pick a number; I pick a number,' he said. 'We multiply them together. If the answer starts with 1, 2 or 3 then I win, but if the answer begins with 4, 5, 6, 7, 8 or 9 you win.'

The game looks like it is weighted in my favour because I have six digits to aim for compared to his three. Yet Ted will win most of the time if he chooses his numbers according to the Benford's percentages. In other words, if he chooses, over the course of several games, a number beginning with a 1 as near as possible to 30.1 per cent of the time, a number beginning with a 2 as near as possible to 17.6 per cent of the time, and so on. If Ted selects in this way, the number I choose makes no difference to the first digit of the answer: 30.1 per cent of the time 1 will be the first digit, 17.6 per cent of the time 2 will be, and 12.5 per cent of the time 3 will be. The sum of these three percentages is 60.2, so Ted will win the game 60.2 per cent of the time. The game is a good one to win money on: with only 1, 2 and 3 as target digits you have far better odds than with 4, 5, 6, 7, 8 and 9, even though it looks like you don't.

The game helps explain why many naturally occurring data sets follow Benford's law. Let's say that Ted and I play a hundred times, and that his numbers are $(a_1, a_2, a_3 \ldots a_{100})$, and that my numbers are $(b_1, b_2, b_3 \ldots b_{100})$. We know that if Ted's numbers are Benford's-compliant then the set of his numbers multiplied by my numbers, or $(a_1 \times b_1, a_2 \times b_2, a_3 \times b_3 \ldots a_{100} \times b_{100})$, is also Benford's-compliant. Consequently, if we multiply these numbers by another set of randomly generated numbers $(c_1, c_2, c_3 \ldots c_{100})$ to get an updated set $(a_1 \times b_1 \times c_1, a_2 \times b_2 \times c_2, a_3 \times b_3 \times c_3 \ldots a_{100} \times b_{100} \times c_{100})$, this will again be Benford's-compliant. The point here is that however many sets of numbers we multiply together in this way, only one of them needs to be Benford's-compliant for the final set of multiplications to be. The law, in other words, is so infectious that only one Benford's set in any multiplicative chain will contaminate the overall result. Since many phenomena– share prices, populations, river lengths, and so on – are made up of increases and decreases caused by many independent, random factors, it becomes less surprising that the lopsided distribution appears so ubiquitously.

Ted's most celebrated theorem states that:

If you take random samples from randomly chosen data sets, then the more sets and samples you select, the closer and closer the first-digit distribution of the combined samples will be to Benford's law.

The theorem tells you when to expect Benford's law. 'If a hypothesis of unbiased random samples from random distributions is reasonable, then the data should follow Benford's law closely,' said Ted. The result explains why newspapers illustrate Benford's law so well. The numbers that appear in the news are effectively random samples taken from random data sets, such as share prices or weather temperatures or voting intentions or lottery numbers. While many of these sets will not follow Benford's law, the more sets we consider and the more samples we include, the closer and closer the combined samples will. If we carry on indefinitely, the combined samples will follow Benford's with a certainty of 100 per cent.

I asked Ted if his theorem had an easy, intuitive explanation. He shook his head. He proved his theorem using ergodic theory, an advanced field that mixes probability theory and statistical physics, and which is only taught at postgraduate level. Despite being straightforward to describe, his theorem has no simple proof. 'Nor is there one in sight. It defies an easy derivation.'

Ted's work, however, has provided the mathematical grounding for the use of Benford's law in legal cases. He has subsequently become the go-to guy for scientists who want to know whether or not they should expect the law to govern their data. One of the oddest requests, he said, came from a Christian group, which had found that the percentages of different minerals in the sea, and in the Earth's crust, all followed the law. This finding was so amazing and so surprising, they said, it could only be the work of an intelligent designer. Would Ted mind testifying as part of their campaign to have creationism taught in Texas schools?

Ted has had fun looking at where Benford's law appears in pure mathematics.

The doubling sequence,
   1, 2, 4, 8, 16, 32, 64, 128, 256, 512, 1024 …

and the tripling sequence,
   1, 3, 9, 27, 81, 243, 729, 2187, 6561, 19683 …

and even the sequence where you switch between doubling and tripling,

1, 2, 6, 12, 36, 72, 216, 432, 1296, 2592, 7776, 15552 …

all obey Benford's law.

So does the Fibonacci sequence, in which each term is the sum of the previous two:

1, 1, 2, 3, 5, 8, 13, 21, 34, 55, 89, 144 …

Which is to say that the more terms you take, the more closely the distributions of the leading digits in the sequence approach Benford's percentages.

Ted has also proved that any sequence starting with a random number and following the rule 'double it and add 1' obeys Benford's law. So does any sequence starting with a random number and following the rule 'square it'. But he discovered something surprising when he looked at the sequence that follows the rule 'square it and add 1'.

'From almost all starting points the sequence follows Benford's law,' he said. 'But there are some points that begin sequences that don't, and they are very hard to find. I didn't think that they existed. I thought "It can't be! It can't be!" But we found one. This number has the amazing property that if you square it and add 1 for ever the first digit is 9. That's amazing. It's a bug in the system.'

The number is 9.94962308959395941218332124109326…

Actually, there are an infinite number of bugs for 'square it and add 1', but they are so thinly spread along the number line that the probability of choosing one randomly is exactly zero. Many aspects of Benford's law, Ted added, remain to be discovered.

Benford's law is one of the most spectacular examples of how a process involving large numbers of unknown random factors can generate a very simple numerical pattern. The exact chain of events that causes a share price to rise and fall, or a town's population to grow, may be too wild or complex ever to be fully understood, yet the outcome is well disciplined and uncomplicated. We may not be able to predict the future price of any individual stock or the population of any individual town, but we can be confident that, overall, stocks and populations will always adhere to Benford's law.

Books also contain simple numerical patterns. Take James Joyce's *Ulysses*. In the 1940s, researchers at the University of Wisconsin spent fourteen months compiling a list of all the words used in it. They typed out the book on gummed material, cut out the individual words and stuck them on tens of thousands of individual sheets of paper. They then ranked the words in order of frequency. The data was of interest not just to language students, but also to psychologists studying word association and to academic mavericks like George Kingsley Zipf, a professor of German at Harvard, who spotted something stunning:

Word	Rank	Frequency
I	10	2653
say	100	265
bag	1000	26
orangefiery	10,000	2

The tenth most popular word was almost exactly ten times more frequent than the hundredth, almost exactly a hundred times more frequent than the thousandth, and almost exactly a thousand times more frequent than the ten thousandth. Joyce had not purposefully chosen his words with such arithmetical precision, yet the pattern jumps off the page.

In mathematical terms, the words in *Ulysses* roughly obey the relationship:

$$\text{frequency} \times \text{rank} = 26{,}500$$

We can rephrase this as:

$$\text{frequency} = \frac{26{,}500}{\text{rank}}$$

Which is an equation of the form:

$$\text{frequency} = \frac{k}{\text{rank}}, \text{ where } k \text{ is a constant}$$

Which is equivalent to saying that frequency is inversely proportional to rank. In other words, if you multiply the rank by an arbitrary number n, you must divide the frequency by n.

After studying other texts, Zipf came to the conclusion that, for all books in all languages, word frequency and rank have an inverse relationship, but with a slight modification:

$$\text{frequency} = \frac{k}{\text{rank}^a}, \text{ where } k \text{ and } a \text{ are constants}$$

The equation is known as Zipf's law. (When two numbers are written in the form x^y, we say 'x to the power y' and it means x multiplied y times. As we know from school, $4^2 = 4 \times 4$, and $2^3 = 2 \times 2 \times 2$. The number y, however, does not need to be a whole number, so $2^{1.5}$ means 2 multiplied by itself 1.5 times, which is 2.83. The closer y is to 1, the closer x^y is to x.)

Zipf discovered that a is always very close to 1, irrespective of writer or subject matter, which means that the relationship between frequency and rank is always very close to one of inverse proportion. In the case of *Ulysses*, a is 1.

I find Zipf's law exhilarating. It reveals not only that there is a mathematical pattern governing word choices, but that this pattern is beguilingly simple. I decided to see if the law holds up for the book you're now reading. To count word frequencies I used computer software, rather than gummed paper and scissors. Scrolling down the table of frequencies, I could see that frequency was indeed roughly inversely proportional to rank. The most common word I use in this book, 'the', appears about ten times more than 'was', the tenth most common word. 'The' appears about a hundred times more than 'who', the hundredth most common word, and it appears about a thousand times more than 'spirals', the thousandth most common word.

When I plot this book's rank/frequency data on a graph,

illustrated below left, the dots look like they are stuck to the axes. Graphs of inversely proportional relationships always produce this type of L-shaped curve. The initial drop-off is drastic, and quickly flattens into a 'long tail', emphasizing that a few words are used in huge numbers, but almost all words are used hardly at all. (In fact, in all texts, whatever their size, about 50 per cent of all words are used only once. In this book, the percentage is 51.)

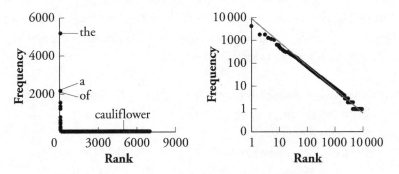

Word-frequency distribution for Alex Through the Looking-Glass.

The right-hand graph above uses the same data but changes the scale. The distances between 1 and 10, 10 and 100, and 100 and 1000 are now equal for both axes, which is known as a double logarithmic, or log-log, scale. Ping! The sagging cable magically transforms into a taut girder. A mathematical order has appeared. The dots are now beautifully arranged in a line.

A straight line on a log-log scale is proof that the data obeys Zipf's law (I explain why in Appendix Two on p. 293). A straight line is more mathematically useful than a drooping long-tailed curve, since its properties are easier to analyse. For example, a straight line has a constant gradient. We will return to the concept of gradient later in this book, but for the time being all you need to know is that the gradient is a 'measure of steepness', the vertical distance covered by the line divided by the horizontal distance covered. If we draw a 'best-fit' line through the dots, and measure its gradient, the gradient is the constant a in the Zipf's law equation. I calculated the gradient a for the line above. It is slightly more than 1, which means that compared to James Joyce I over-use my most frequent words, and under-use my least frequent ones.

On closer inspection, the dots are not all exactly on a straight line. Some of them diverge from the best-fit line, especially those representing the 20 or so highest-ranked words. Overall, however, the dots behave in a remarkably tidy way. It is astonishing that for the vast majority of words in this book, their rank determines pretty accurately the number of times they are used, and vice versa.

Professor Zipf discovered the same inversely proportional relationship in another tome, the 1940 United States Census. However, this time he wasn't counting word frequencies but looking at the populations of America's largest cities.

Metropolitan district	Rank	Population
New York/NE New Jersey	1	12m
Cleveland	10	1.2m
Hamilton/Middletown	100	0.11m

Again, the pattern is almost too good to be true. New York, the largest city, has ten times the population of Cleveland, the tenth largest city, and a hundred times the population of Hamilton, the hundredth largest city. No one told Americans to confer or acquiesce so precisely in their choice of address. Yet when deciding where to live they did so in a tightly regulated fashion. They still do. In fact, we all do. The graphs opposite show the population/rank data for American cities from the 2000 census, and for the 100 largest countries in the world, on log-log scales.

Like obedient ants, the dots fall in a straight line, which means that the same general equation as before applies:

$$\text{population} = \frac{k}{\text{rank}^a}, \text{ where } k \text{ and } a \text{ are constants}$$

Again, Zipf found that for cities and countries, a is close or equal to 1. For American cities it is 0.947 and for the world it is 1.156. For the 1940 census, a is 1.

Of course, there are wobbles, especially for the largest countries and cities. The second most populous country in the world, India,

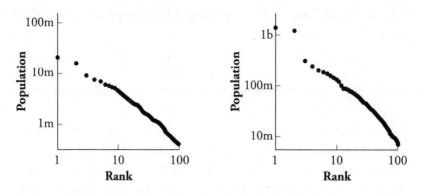

Population distributions of the largest cities in the US in 2000 (left), and the largest countries in the world in 2013 (right).

has more inhabitants than Zipf's law would suggest. But the volatility at the beginning of the rankings is inevitable since there are far fewer data points. One would expect cities and countries to overtake each other as their populations change due to economic, social and environmental factors. When this change happens to the highest-ranked countries, the deviation from the straight line is more noticeable. Still, this early scatter should not detract from the tightness of fit further down the line. The behaviour of words, cities and countries appears to conform to a universal law.

Finding the same, elementary mathematical pattern in different contexts was for Zipf a spiritual awakening. 'We are finding in the everyday phenomena of life a unity and orderliness and balance that can only give faith in the ultimate reasonableness of the whole whose totality lies beyond our powers of comprehension,' he wrote, and proposed a 'principle of least effort' as a theoretical basis for his empirical observations. We use a few words a lot because it is easier on our brains and we live in large cities because it is more convenient. Zipf, however, failed to come up with a proper mathematical derivation for his law, and after almost a century no one else has either. Many people have tried, and while a few have claimed qualified success in some areas, the reason why the law is true is still a mystery. Often mathematical models are criticized because they oversimplify complex behaviour. In the case of Zipf the opposite is true: the models are impossibly difficult, but the pattern is so straightforward a child can understand it.

In the early 1900s, the Italian economist Vilfredo Pareto stated that the distribution of wealth among a population follows this law:

$$\text{the wealth of an individual} = \frac{k}{\text{rank}^a} \text{ , where } k \text{ and } a \text{ are constants}$$

We instantly recognize Pareto's law as mathematically equivalent to Zipf's. If you rank everyone in a country according to their wealth, the graph of how wealth is spread looks just like the graph of the frequency of words in this book on p. 41. On the whole, the richest person in a country is quite a bit richer than the second richest, who is quite a bit richer (but less so) than the third richest, who is a bit richer (but less so) than the fourth richest, and so on. Taken overall, only a tiny minority of the population is rich and most people are poor. Pareto found his law in data from many countries and across many eras, and to a large extent it still applies today.

Inversely proportional relationships describe situations of extreme, outrageous inequality. In the case of Zipf's law, a tiny percentage of words do almost all the work. In the case of Pareto's law, a small percentage of the population owns most of the wealth. In 1906, Pareto wrote that in Italy about 20 per cent of the people owned 80 per cent of the land. His zinger entered popular culture as the Pareto principle, or the 80/20 law: the rule that 20 per cent of causes produce 80 per cent of effects, a slogan about the unfairness of life. According to Richard Koch, who wrote a book about the Pareto principle, 20 per cent of employees are responsible for 80 per cent of output, 20 per cent of customers bring in 80 per cent of revenue, and we achieve 80 per cent of our happiness during 20 per cent of our time. The 80/20 law is the 'key to controlling our lives', he writes, because only by focusing on the vital 20 per cents can we transcend the pressures of the modern world. The Pareto principle is memorable because it is arithmetically neat that 80 + 20 = 100, but this neatness is largely irrelevant to the mathematical pattern being expressed, which is that many things are roughly inversely proportional to each other.

Pareto and Zipf's laws both state that one quantity is inversely proportional to the *power* of another quantity.

If the variable quantities are x and y, then the general form of this mathematical statement is:

$$y = \frac{k}{x^a}, \text{ where } k \text{ and } a \text{ are constants}$$

Equations of this form are called 'power laws'. Zipf and Pareto gave their names to the two most famous ones, but in recent years power laws have been found in an extraordinary variety of contexts. For example, a Swedish survey of sexual habits found that:

$$\text{percentage of men with at least } n \text{ sexual partners over the last year} \approx \frac{k}{n^{2.31}}$$

The symbol $\approx$ is not a comment that Swedish women prefer men with wavy moustaches. It means 'roughly equal to', and is used because the equation is an approximation of the best fit. About one in a thousand Swedish men have twenty partners a year, but most have only one. If we extend the best-fit line on the graph overleaf, the survey suggests that about one in ten thousand men will have about sixty partners a year.

As in love, so in war. Researchers studying violence in conflict zones discovered that:

$$\text{percentage of incidents in the Colombian civil war with a number of deaths and injuries of at least } n \approx \frac{k}{n^{2.5}}$$

Incidents with lots of deaths and injuries are extremely rare, compared to the many incidents with only one casualty. Similar results have been shown for other wars, and also when comparing wars. There are only a few wars in which millions of people died; there are slightly more in which hundreds of thousands died; more still in which tens of thousands died, and so on.

Charles Darwin sent thousands of letters in his lifetime, many of them replies to letters he received. He responded to most of them in one day, but some took him years:

$$\text{probability that Charles Darwin will respond to a letter in } n \text{ days} \approx \frac{k}{n^{1.5}}$$

We answer our emails following the same pattern: most we reply to instantly, but some linger for an eternity at the bottom of the inbox.

Japanese academics looked at the volume of book sales between 2005 and 2006. They found that:

$$\text{percentage of total sales in Japan in 2005/2006 of book ranked } n \approx \frac{k}{n^{0.65}}$$

In other words, a few books are bestsellers but most go unsold. The film industry's business model is based on this pattern: a minority are blockbusters, and they pay for the majority that are turkeys. In both cases the slide from success to failure is mathematically predictable.

The four equations above were deduced by plotting the experimental data on log-log scales, illustrated opposite, and measuring the gradients of the best-fit lines. (The droop at the end of the Japanese data is explained by limits on shelf space: bookstores are not big enough to hold all the books that theoretically would have to be in stock.) A straight line on a log-log graph means you have a power law, and the gradient of the line is the constant a in the power-law equation. I left out the values for the constant k in each equation because, as an indicator of the size of the sample, not the shape of the curve, it is not interesting to us. Remember, if the data in each case had been plotted on normal scales it would make the shape of an L, with a steep initial drop and a long tail.

My purpose in presenting so many examples is to make you see the world more like George Zipf, Vilfredo Pareto and Richard Koch. When we consider, say, heights in a population, there is

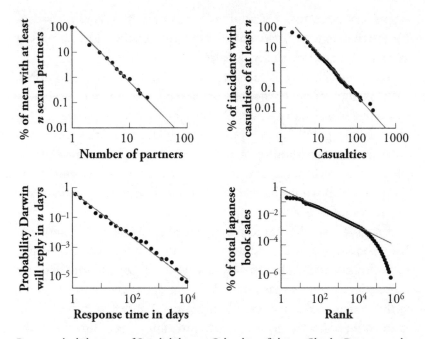

Data on the behaviour of Swedish lovers, Colombian fighters, Charles Darwin and Japanese book buyers reveals power laws.

an average height. The numbers cluster around a middle value, such as 5ft 9ins for British men. But when we talk about word usage, wealth, sexual partners, wars, letter response times, books and films, we cannot talk about averages in the same way. Just as there is no averagely used word, or averagely wealthy person, there is no averagely selling book or averagely grossing film. When it comes to human behaviour, we live in a world weighted towards extremes.

Not only are power laws ubiquitous in the human sciences, they also abound in the physical sciences. The magnitude of an earthquake is roughly inversely proportional to the number of earthquakes of that magnitude; the size of a moon crater is roughly inversely proportional to the number of craters of that size; and if you fire a frozen potato against a wall, the size of each fragment is inversely proportional to the number of fragments of that size. The prevalence of power laws in physics explains why many scientists

47

who study them in social systems began their careers as physicists. One such person is Albert-László Barabási, a distinguished professor at Northeastern University in Boston.

Barabási's current field is networks, and in certain networks, such as the internet, there is an accepted mathematical theory for why power laws emerge. The popularity of websites, for example, roughly follows a power law, as does, say, the ranking of Twitter users by number of followers. 'That [power laws] are so generic, universal and unmistakably *there* is very puzzling,' said Barabási. 'You would have thought there would be more diversity in the world!'

Imagine that the figure below left is the model of a network. It is made up of three nodes and two links. Nodes can be people, or websites, and links can be any type of connection between them. Barabási says that power laws emerge if the network grows with 'preferential attachment', meaning that when a new node is introduced to the network, the probability of a link from the new node to any node already in the network is proportional to the number of links the node already has. In other words, well-connected nodes become even better connected. The rich get richer. The famous get more famous. The node with the most links has the highest chance of getting new links, and the more links it gets the more attractive it becomes.

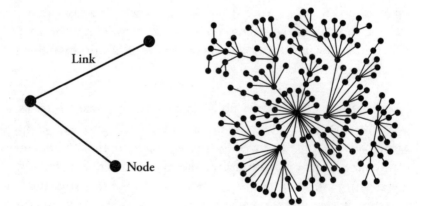

If it grows by preferential attachment, the small network will eventually resemble the large one.

If the network on the left was allowed to grow by preferential attachment, after a couple of hundred nodes were added it would end up looking something like the network on the right. Most nodes have only one link, and only a few nodes, the hubs, have several links. If you ranked the nodes by their number of links, and plotted the data on a graph, you would see the familiar long tail curve. 'Every time you give the decision [about who to link] to the node, a power law will emerge,' Barabási says. If you added several million more nodes by preferential attachment, the network would begin to look like a map of who's following who on Twitter, or a model of the World Wide Web.

One reason power law networks are so common, adds Barabási, is that they are particularly robust. If you delete a node randomly, you are much more likely to delete an unimportant node – since there are many more of them – than a hub, so the overall effect on the network will be minimal. Although, conversely, this makes power law networks especially vulnerable if a hub is singled out for attack. If my website crashes, no one cares apart from me, but when Google goes down for five minutes there's global pandemonium.

Power laws are appealing because they provide a startlingly simple mathematical model for an array of complex phenomena. They are also seductive because they are so easy to detect. As we have seen, two variables follow a power law when the data points fall in a straight line on log-log scales.

In recent years, however, it has been suggested that scientists can be too quick to claim power laws in their data, since sometimes the dots fall in curves that vary slightly from a straight line, and which are described by different equations. It's an important debate, but not one for this book. One aspect of power laws that cannot be denied, however, is that they exhibit the following fascinating mathematical property.

Let's consider the power-law equation $y = \frac{1}{x^2}$. When we plot the graph of this equation for x between 2 and 10 we get the curve on the left overleaf, and when we plot the equation between 20 and 100 we get the curve on the right.

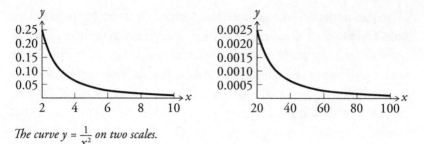

The curve $y = \frac{1}{x^2}$ on two scales.

Spot the difference? The curves are exactly the same. In fact, when we plot the curve between n and $5n$ for any number n, the curve will look the same as these two. Likewise, the curve always looks the same between any two numbers a and b when the ratio a/b is fixed. Power laws exhibit the same pattern on all scales, however far you voyage down the tail.

Speaking of long tails, Godzilla had one.

The Japanese monster, a kind of mutant dinosaur, was also supposed to be about 100m high, which is about 50 times the height of a tall adult man. Now imagine enlarging a man so he is 50 times higher, but still the same shape. The enlarged man will be 50 times as wide and 50 times as thick, and therefore $50 \times 50 \times 50 = 125,000$ times as heavy as he was before. The cross section of his bones, however, will only have grown $50 \times 50 = 2,500$ times, meaning that each square inch of his bones has to support fifty times as much weight. The giant man's bones would snap the moment he tried to walk. Godzilla would suffer a similar fate.

Okay, there's nothing more tedious than a smart alec moaning that a movie monster couldn't really exist. Yet the argument also explains why animals of different sizes tend to have different shapes. The bigger an animal is, the thicker its bones must be relative to its height, an observation first made by Galileo in 1638. An elephant has proportionately thicker bones than a human, which has proportionately thicker bones than a dog. The bones are thicker, Galileo realized, since they need to withstand more weight per cross-sectional area.

We can translate Galileo's observation about bone shape into an equation involving area and volume. The statement that the cross-sectional area of an object increases proportionately with the *square* of the height, while the volume increases proportionately with the *cube* of the height, can be rewritten by the two equations:

$$\text{area} = l \, (\text{height})^2$$

and

$$\text{volume} = m \, (\text{height})^3$$

where l and m are constants

We can eliminate the variable height to get the equation:

$$\text{area}^3 = \frac{l}{m} \, (\text{volume})^2$$

Which rearranges to:

$$\text{area} = \frac{l}{m} \, (\text{volume})^{2/3}$$

This is an equation of the form:

$$y = kx^a$$

where x and y are variables, and k and a are constants

This type of equation is also called a power law. When a power law takes this form, we say that y is *directly* proportional to x^a, whereas when a power law is of the form discussed previously, $y = \frac{k}{x^a}$, the variable y is *inversely* proportional to x^a.

The graph of the power-law equation $y = x^{2/3}$ is illustrated overleaf. On the left graph, which has normal scales, the curve flattens out as it rises. If we think of y as area, and x as volume, then this shows that as the volume increases, the area also increases, but not as fast. A directly proportional power law produces a straight line on a log-log plot, also illustrated overleaf, in which the slope leans to the right.

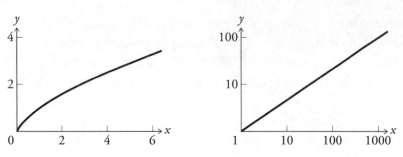

The curve y = x^⅔ on standard and double logarithmic scales.

The power-law equation between volume and area is also called a 'scaling law', because it shows what happens to a measurable quantity of an object, in this case cross-sectional area, as the object increases in overall size.

In the 1930s, the Swiss zoologist Max Kleiber measured the weights of several different species of mammal, and their metabolic rates, which are the rates at which the animals produce energy while at rest. When he plotted the data on a log-log graph, there was a straight line, from which he deduced the following power law:

$$\text{metabolic rate} \approx 70 \ (\text{mass})^{\frac{3}{4}}$$

It is known as Kleiber's law, and biologists have subsequently extended it to all warm-blooded animals, as illustrated opposite. Metabolic rate does not rise as fast as mass, showing that animals get more efficient at producing energy as they get bigger. Many other scaling laws have been found in animals: lifespan is directly proportional to $(\text{mass})^{\frac{1}{4}}$ and heart rate is inversely proportional to $(\text{mass})^{\frac{1}{4}}$. Since the power-law coefficient seems always to be a multiple of $\frac{1}{4}$, biological power laws are called 'quarter-power-scaling laws'. Considering the diversity of the animal kingdom – mammals vary in size from the Etruscan shrew, at around 1g, to the blue whale, which is 100 million times heavier – it is remarkable to think that once you know the size of an animal, you can already predict so much about it.

The physicist Geoffrey West, of the Santa Fe Institute, and the biologists James Brown and Brian Enquist of the University

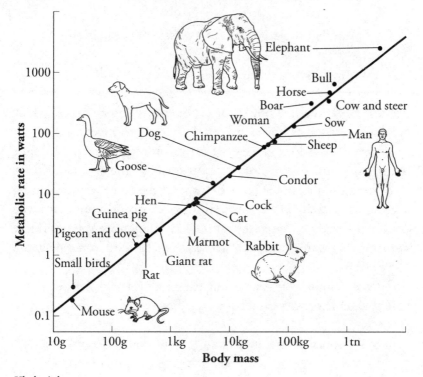

Kleiber's law.

of New Mexico, have devised a mathematical theory that explains quarter-power scaling. In broad terms, they argue that if you treat an organism as a transport mechanism – blood travels down the aorta, which branches into arteries, which themselves branch into narrower vessels – when you optimize the system to fit the space, you get a power law. The details are beyond the scope of this book, but are of interest here because of how they influenced West's follow-up work: studying another type of organism, the city.

West and his colleagues discovered that towns and cities are strongholds of power-law scaling. After crunching vast amounts of economic and social data, and plotting many log-log graphs, they found that in the US, for example:

$$\text{number of inventors} \approx k \, (\text{size of population})^{1.25}$$

$$\text{total wages} \approx k \, (\text{size of population})^{1.12}$$

$$\text{number of Aids cases} \quad \approx \quad k \text{ (size of population)}^{1.23}$$

$$\text{number of serious crimes} \quad \approx \quad k \text{ (size of population)}^{1.16}$$

In these equations the power, or exponent, is larger than 1, which means that as a city gets bigger, it has more inventors, wages, Aids cases and serious crimes per capita. You get more in absolute terms, but you also get more proportionately. The exponent for all these urban indicators is around 1.2, and the fact that they cluster together is in itself curious. With this figure, it works out that if you double the size of a city you can expect to increase the per capita number of inventors, wages, Aids cases and serious crimes by about 15 per cent.

For some indicators, the scaling exponent is less than 1, meaning that you get fewer per capita as a city grows:

$$\text{number of petrol stations} \quad \approx \quad k \text{ (size of population)}^{0.77}$$

$$\text{length of electrical cables} \quad \approx \quad k \text{ (size of population)}^{0.83}$$

If you double the size of a city, you can expect a per capita reduction of about 15 per cent in the number of petrol stations and electrical cables. In other words, cities show mathematically predictable economies of scale, and they do so all over the world. 'The evolution of Japanese cities was completely independent of the evolution of European or American cities, yet the same scaling turned up [in each country],' says West. 'That is suggestive that there is some universal dynamic at work.' West believes that power laws appear in cities for the same reason that he argues they do in animals. The city is also a transport network. Just as the circulatory system transports blood down a thick tube into thinner and thinner ones, so cities distribute resources down networks of branching roads, cables and pipes.

Where we choose to live is our own decision, as is how we spend our money and what we do with our time. Yet seen through the lens of numbers our collective behaviour is predictable, and obeys simple, mutually compatible mathematical laws. We have spread ourselves across the globe in such a way that in every country about 30 per cent of towns and cities have a population that begins with a 1, city size is roughly inversely proportional with rank, and all cities are just power-law-scaled versions of each other. The world may be complicated. It is also simple.

Numbers are indispensable tools that help us understand our surroundings. So are shapes. It was thanks to the study of one shape in particular that Western mathematics began.

Love Triangles

Rob Woodall is a trig bagger. He bags trigs, and no one has bagged as many as him. Trigs are trig pillars, waist-high concrete columns that mark out a national grid of reference points once used by mapmakers and surveyors. You've probably passed one if you've ever walked in the British countryside. They are usually on the tops of hills, the prize at the end of a climb. The Ordnance Survey built more than 6500 trigs between 1936 and 1962, of which about 6200 survive. Visiting, or 'bagging', as many as possible is a competitive sport. Rob, who is 50, has bagged 6155, almost the full set. He is currently about a thousand ahead of his nearest rival.

When he started, Rob would spend every other weekend on missions, leaving his home in Peterborough on Friday night and returning on Monday morning. Trigs are spaced on average about 5km apart, and working swiftly he aimed for a rate of about 50 each weekend. If he was lucky the pillars were by the side of the road, where he could park his car, but often they were far from roads or footpaths, hidden in the middle of gorse, brambles or thorny hedges. He started to travel with a pair of secateurs to avoid arriving at work with bloody hands.

Trig pillars are relics of our technological heritage, archaeological features of our landscape like medieval forts or straight Roman roads. Rob likes bagging them because it takes him to attractive places and gives him a sense of adventure and achievement. He has made night-time dashes across farmers' fields, he once jumped into an ostrich pen, and he spent three years negotiating with one particular landowner before gaining permission to see a pillar on his property. I love trig pillars too. They are shrines to the majesty of the triangle, the shape that changed the world.

———

Numbers emerged about 8000 years ago but mathematics proper arrived in Egypt around 600 BCE.

It began with a public demonstration. Thales, a Greek thinker, showed how to measure the height of the Great Pyramid at Giza without having to climb it. He first placed a staff on the ground, which, together with its shadow, created two sides of a triangle, as illustrated below. The pyramid and its shadow also created a triangle. Thales's genius was to realize that even though the two triangles were of vastly different sizes, they had the same shape, because the sun's rays are parallel. This observation meant he could use the height of the small triangle to deduce the height of the large one. In modern terms, he understood that:

$$\frac{\text{height of staff}}{\text{length of staff shadow}} = \frac{\text{height of pyramid}}{\substack{\text{distance from centre of pyramid base} \\ \text{to tip of shadow}}}$$

The staff and its shadow are easily measured. The distance along the ground from the pyramid base cannot be measured directly, since the pyramid is in the way; Thales may have waited until the sun was perpendicular to the pyramid before taking measurements, because at that time the distance from the centre of the pyramid to its edge equals half the length of its side. With three of the values in the equation ascertained, he was able to calculate the remaining value – pyramid height.

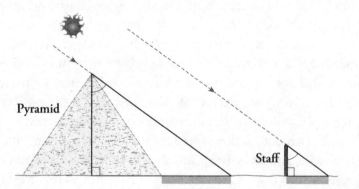

The sun's parallel rays create two similar triangles, one cast by the pyramid and one by the staff.

Thales's achievement was a small step for trigonometry, the science of triangles, and a giant leap for mankind. By deducing a measurement logically from the intrinsic properties of a shape he was thinking differently from the Egyptians, who had shown remarkable skills in practical activities like pyramid-building but whose mathematical knowledge was essentially limited to rules of thumb and triangles that actually existed. Thales's calculation involved a triangle that was an abstraction of reality, made by the sun's rays. His insights marked the beginning of Greek rational thought, which we generally consider the foundation of Western mathematics, philosophy and science.

Thales is also the first person to have a specific mathematical discovery named after him: Thales's theorem, which states that the triangle inscribed in a semicircle has a right angle. He also used his deductive powers to predict the solar eclipse of 585 BCE, and to predict that the olive harvest in his home town of Miletus would improve after a few bad years. He bought all the olive presses he could at rock bottom prices and when the upturn came he got rich. A century later, the comic playwright Aristophanes made fun of the great sage by having him fall in a ditch because he was lost in thought, gazing at the sky. Thales is not just remembered as history's first mathematician and philosopher, but also as history's first absent-minded professor.

Thales's performance at Giza showed how the triangle could be used to measure the distance from a near point to a far point that you physically didn't need to reach. Triangles would later be used to measure much larger distances than the height of a pyramid, thus transforming the sciences of astronomy, navigation and cartography. We will get there shortly. Sometimes, however, enormous distances could be measured simply by contemplating the shadow cast by a vertical staff on a sunny day. Three centuries after Thales used shadows and deductive logic to impress the Pharaoh, Eratosthenes employed the same technique to make the first realistic estimate of the size of the Earth.

Eratosthenes lived in Alexandria, capital of Hellenistic Egypt, where he was in charge of its library, the largest in the Greek empire.

At noon on midsummer's day in Alexandria, he measured the 'shadow angle' at the tip of a vertical staff to be about a fiftieth of a full circle. He also knew that in Syene, the most southerly city in Egypt, there was a famous well whose bottom lit up completely at noon on midsummer's day, meaning that the sun cast no shadow at all at that specific time and place. From these two facts he deduced that the distance from Alexandria to Syene must be a fiftieth of the circumference of the Earth.

His reasoning went as follows. First, it was known by then that the world was round – people had observed ships sinking behind the horizon and seen the Earth casting a curved shadow on the moon during a lunar eclipse. Secondly, Eratosthenes knew that Syene was roughly due south of Alexandria. From this he was able to draw the illustration below left, which shows a north-south cross section of the Earth through Alexandria and Syene at noon on midsummer's day. The sun's rays go straight down the Syene well in the direction of the centre of the Earth, and hit the Alexandria staff at an angle. Since the staff is positioned vertically, it must also point to the centre of the Earth, and from this we can make the geometrical abstraction illustrated below right, where the parallel lines represent the sun's rays and the line crossing them is the line from the tip of the staff to the centre of the Earth.

A basic theorem of Greek geometry is that 'alternate' angles are equal, meaning that a line falling on two parallel lines crosses them both at the same angle, hence the angle cast by the staff is equal

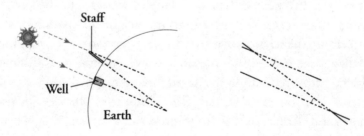

At noon on midsummer's day, the sun casts no shadow down the well at Syene, but casts a shadow from the staff in Alexandria. The staff's 'shadow angle' is equal to the angle between the two cities at the centre of the Earth.

to the angle at the centre of the Earth. Eratosthenes measured the angle cast by the staff to be a fiftieth of a circle, and so deduced the angle at the centre of the Earth to be the same. It follows that the distance between Alexandria and Syene is a fiftieth of the distance all the way around the world.

So, in order to estimate the circumference of the Earth, Eratosthenes simply needed to multiply the distance between Alexandria and Syene by fifty. The Greeks already had a pretty good estimate for this distance, 5000 stades, which had been calculated by *bematists*, or step-people, whose job it was to walk for days on end, counting their strides. (Eratosthenes, who invented geography, was blessed by three pieces of geographical serendipity without which his measurement would have been impossible: that the Egyptians had settled as far south as Syene, on the Tropic of Cancer, the northernmost latitude where the sun casts no shadow at least once a year; that Syene was due south of Alexandria; and that the terrain between them enabled the road to proceed more or less as the crow flies.) A stade is estimated to be about 166m in modern measure, so the calculation for the circumference of the Earth was 166m × 5000 × 50, which works out at 41,500km, only 1500km – about 4 per cent – more than the correct value. No one improved on Eratosthenes's result for almost a thousand years.

Syene is now known as Aswan, and it even has a well you can visit, although the brutal noonday heat on the summer solstice means it is unlikely to become a tourist attraction.

By Eratosthenes's time, Greek mathematics had expanded from Thales's early insights about triangles into a sizeable corpus of theorems and proofs about triangles. The primacy of the triangle in Greek thought came about because shapes made out of straight lines – squares, pentagons, and so on – can all be divided into triangles, and shapes with non-straight lines like circles, ellipses and parabolas can all be approximated by triangles.

In fact, since all triangles can be divided into *right-angled* triangles – which are those that contain a right angle, or 'quarter-turn' – the Greeks prized right-angled triangles above all others. The illustration on the left overleaf shows how to split a triangle into two smaller ones with right angles: draw a perpendicular line from the largest side to the opposite corner. We are taught about right-angled triangles early in our mathematics education, learning the word 'hypotenuse', the side opposite the right angle, and, immediately afterwards, Pythagoras's Theorem, illustrated below right, which says that:

> *For right-angled triangles, the square of the hypotenuse is equal to the sum of the squares of the other two sides.*

Pythagoras's Theorem is the most famous theorem in mathematics for many reasons, the most important of which is because it concerns the right-angled triangle, the irreducible unit of plane geometry.

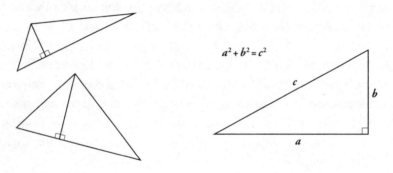

$$a^2 + b^2 = c^2$$

Right-angled triangles.   *Pythagoras's Theorem.*

When the sun casts a shadow on a staff, it makes a right-angled triangle, as we saw with Thales. But when the sun moves across the sky, a change in angle does not cause a proportional change in shadow length. When the angle increases by a constant increment, as in the illustration opposite, shadow length increases by increments that grow larger each time, which is why at the end of the day we can see shadows creep across the ground. Astronomers, not to mention the makers of sundials, were especially interested in understanding the relationship between shadow angle and shadow length. The Greeks, however, were ill-equipped to answer this question because, for all their geometrical expertise, they had a terribly cumbersome number system. In order for their pursuit of triangle knowledge to advance, they needed a better way of writing fractions.

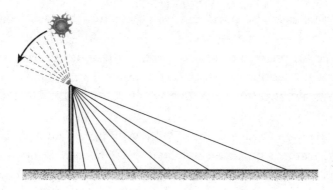

Equal angles cast unequal shadow lengths.

Greek numerical notation was descended from Egyptian notation, which came in two types. When carving on wood or stone, Egyptians used hieroglyphs. Each power of ten from one to a million had its own symbol: a vertical stroke for 1, an inverted U for 10, a spiral for 100, a lotus flower on its stem for 1000, a slightly bent raised finger for 10,000, a tadpole for 100,000, and a kneeling man raising his head to the sky for 1,000,000. A number was expressed through repetition of these symbols, so 3,141,592 was:

When writing on papyrus, they used a less elaborate 'hieratic' script, which was more suited to writing with pen and ink. They introduced special signs to represent digits and multiples of ten, so rather than tediously writing out, for example, seven strokes for 7, they used the single symbol **Ⴭ**. The change from repetition-based numerals to symbol-based ones was an important advance.

Fractions in hieroglyphs were illustrated by a mouth, ⬬ , above a number, signifying its reciprocal, just as we use a 1 above the line in a fraction, so $\frac{1}{3}$ was ⍟ and $\frac{1}{10}$ was ⍟ . Fractions in hieratic script used a dot above a number, so $\frac{1}{7}$ was **Ⴭ**. The Egyptians almost exclusively used unit fractions, which are those with a 1 above the line. The fraction $\frac{2}{5}$ was therefore laboriously broken down into $\frac{1}{3} + \frac{1}{15}$, and $\frac{2}{101}$ was $\frac{1}{101} + \frac{1}{202} + \frac{1}{303} + \frac{1}{606}$. The shrinking values of Egyptian sums of unit fractions hints at our own system of decimal fractions, in which, for example, 0.234 stands for $\frac{2}{10} + \frac{3}{100} + \frac{4}{1000}$, although their system was not as efficient and versatile as ours is.

By the time of Euclid, the Greeks were using a number system derived from Egyptian hieratic script: 27 distinct numbers were represented by 27 distinct symbols, the letters of the Greek alphabet. The number 444 was written υμδ, because υ was 400, μ was 40 and δ was 4. Fractions were described rhetorically, for example as 'eleven parts in eighty-three', or written as common fractions with a numerator and a denominator, much like the modern form $\frac{11}{83}$, although the Greeks maintained the historic obsession with unit fractions. The Egyptian and Greek systems were ill-suited to astronomy, which requires tiny subdivisions of angles to track the movements of planets, because common and unit fractions are so unwieldy.

In Mesopotamia, however, a much more flexible numerical notation was in use. The Babylonians used a 'place value' number system, in which the value of a number depends on its position. Our modern number system is a decimal place value system: the number 123, for example, means 3 in the units column, 2 in the tens column and 1 in the hundreds column. A great advantage of a place value system is that it can be expanded to cover fractions. In our notation, we call this type a 'decimal' fraction. The number 0.56 means 5 in the tenths column and 6 in the hundredths column.

The Babylonians used a 'sexagesimal' place value system, meaning that they counted in sixties instead of tens: the number written '123' meant 3 in the units column, 2 in the 60s column and 1 in the $(60 \times 60) = 3600$s column. (Babylonian numerals were composed of combinations of two symbols, a vertical wedge 𒐕 and a horizontal wedge 𒌋). Why they chose sixty as their base is unknown, although it may be because sixty is the lowest number divisible by 1, 2, 3, 4, 5 and 6, which would have made some arithmetical tasks easier. The Babylonians extended their system to fractions. They had no 'sexagesimal' marker like our decimal point, so the value of the columns had to be deduced from the context. The number '123' could therefore also mean 1 in the units column, 2 in the column for sixtieths, and 3 in the column for 3600ths. Positional fractions are vastly superior to common ones, as we know from our experience with decimal fractions. They require fewer symbols, and they are simpler to calculate with. The Babylonians were able to calculate $\sqrt{2}$ to three sexagesimal places, or about 0.000008 from its true value, an amazing result for the period. The ease with which they could subdivide angles helped their astronomy become the most advanced of its time.

The Babylonians divided the circle into 360 degrees. This division may have been influenced by the zodiac, which was divided into 12 signs and 36 decans, or because 360 is roughly the number of days in a year. More recently it has been argued that 360 was chosen because six equilateral triangles fit snugly within a circle, as shown below, and that each of these angles was divided into 60 as demanded by sexagesimal fractions. Certainly, all these reasons complemented each other, and the Babylonian system has proved remarkably durable.

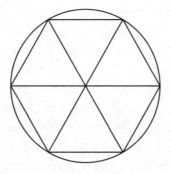

In the second century BCE the Greeks appropriated Babylonian fractions, which have been in use ever since. The degree was traditionally divided into sixty smaller units, each a *pars minuta prima*, or first minute part, which were then divided into sixty smaller units, each a *pars minuta secunda*, or second minute part. From the translation of these Latin phrases we get the words *minute*

and *second*, our units of time, which are the most prominent modern relics of the ancient practice of counting in groups of sixty.

With an appropriate number system finally available to him, the Greek astronomer Hipparchus embarked on a project to chart the relationship between the angles and the sides of a triangle. He did this by considering the 'chord', which is a line between two points on a circle, so called because it looks like a taut string on a bow. Every chord forms a triangle with the centre of the circle, as illustrated below.

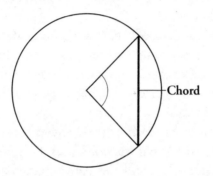

If the size of the circle is fixed, then for each angle at the centre, the chord has a different length. Hipparchus compiled a table of angles that were multiples of 7.5 degrees, together with their corresponding chord lengths. In the second century CE, the astronomer Ptolemy expanded on this idea. He compiled a table of chords for a circle of radius 60 units that listed the angle every half degree from 0 to 180 degrees, together with the length of chord to three sexagesimal places. Hipparchus's and Ptolemy's tables of chords were invaluable to Western astronomers, who calculated distances by considering the Earth and the heavenly bodies as the vertices of interplanetary triangles. The triangle was humankind's earliest telescope, bringing extra-terrestrial locations within the realm of measurement for the first time.

Astronomy flourished in India in the middle of the first millennium CE for the same reason that it had in Babylon: the Indians had a place value number system, which allowed them to describe very large and very small numbers efficiently. The Indian system, in fact, was superior to the Babylonian one because it was based on tens, a more manageable grouping than sixties, and because it treated zero as a fully-fledged number, rather than just a place holder, as the Babylonian system did. Indian astronomers also relied on tables of triangle lengths. Instead of using chords, however, they compiled tables of 'half-chords'. The half-chord, illustrated below left, is also the side of a right-angled triangle where the radius is the hypotenuse, and the other side is part of the perpendicular bisector of the chord. Half-chords are a more convenient concept for calculations since, as we saw before, all triangles can be reduced to right-angled triangles. The Indians' place value system of numbers, and their taste in triangle lengths, spread to the Arab world, and eventually reached Europe. The numbers we use today – the digits 0 to 9 – originate from the Indian system, as does the modern preference for half-chords.

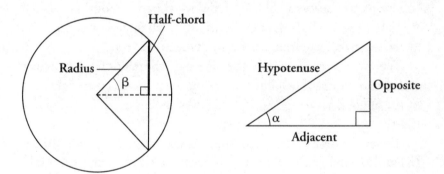

In the sixth century BCE, Thales had already grasped the most important property of triangles, which underpins everything else we know about them: that if the angles are the same, the relative proportions of the sides are the same.

Fast forward two thousand years, and mathematicians invented three new concepts based on this property:

SOH–CAH–TOA!

For those who have forgotten the classroom battle-cry:

$$\text{Sine} = \frac{\textbf{O}\text{pposite}}{\textbf{H}\text{ypotenuse}} \qquad \text{Cosine} = \frac{\textbf{A}\text{djacent}}{\textbf{H}\text{ypotenuse}} \qquad \text{Tangent} = \frac{\textbf{O}\text{pposite}}{\textbf{A}\text{djacent}}$$

The sine, cosine and tangent are called 'trigonometric ratios' and they apply to right-angled triangles, such as the one illustrated on the right on the previous page. The definitions mean that the sine of the angle α is the length of the opposite side divided by the length of the hypotenuse; the cosine of α is the length of the adjacent side divided by the length of the hypotenuse; and the tangent of α is the opposite length divided by the adjacent.

If I were to enlarge the right-angled triangle in the illustration to whatever size I like, the proportions between the sides will stay the same, which means that the sine, cosine and tangent of α, which we write 'sin α', 'cos α' and 'tan α', are always constant. The trigonometric ratios are like ID codes for the shapes of right-angled triangles: the shape is defined by the internal angles, and once the angles are fixed, the sine, cosine and tangent are too.

The connection between the sine and the half-chord becomes clear by looking at both illustrations on the previous page. The sine of the angle β is the $\frac{\text{opposite}}{\text{hypotenuse}}$, which is the $\frac{\text{half-chord}}{\text{radius}}$. If the radius is 1 the sine of β *is* the half-chord.

The etymology of the word 'sine' explains its passage from India. In Sanskrit, the half-chord was *jya-ardha*, or 'string-half'. The Arabs transliterated this as *jiba*, a meaningless term, but it sounds a bit like *jaib*, meaning bosom, or bay, which they came to use. Latin versions of Arab texts translated *jaib* as *sinus,* the word for the fold of a toga over a woman's breasts. *Sinus* became sine.

Here's a small trigonometric table. Clean angles do not always give clean ratios. Between 0° and 90°, the sine goes from 0 to 1, the cosine from 1 to 0 and the tangent from zero to infinity. During the fifteenth and sixteenth centuries, the first trig tables were compiled, using geometrical and arithmetical techniques, setting the stage for the triangle's golden age.

sin 1° = 0.0175	cos 1° = 0.9998	tan 1° = 0.0175
sin 10° = 0.1736	cos 10° = 0.9848	tan 10° = 0.1763
sin 30° = 0.5000	cos 30° = 0.8660	tan 30° = 0.5774
sin 45° = 0.7071	cos 45° = 0.7071	tan 45° = 1.0000
sin 60° = 0.8660	cos 60° = 0.5000	tan 60° = 1.7321
sin 90° = 1.0000	cos 90° = 0.0000	tan 90° = ∞

With the technical stuff out of the way, we can put our new tools to use. If we want to measure the height, say, of a tree, we turn the problem into one involving a right-angled triangle, as shown below.

If P is a point on the ground with a view of the top of the tree, and α the angle of view, then:

$$\tan \alpha = \frac{\text{height of tree}}{\text{distance from viewpoint to tree}} = \frac{h}{d}$$

Which we can rearrange to:

$h = d \times \tan \alpha$

Which is usually written:

$h = d \tan \alpha$

A Renaissance surveyor would have been able to measure α with a protractor and a viewfinder, and once he had α he would have consulted his trigonometric tables to find tan α. He could measure d with a tape measure or a piece of string. Voilà: a way to measure tree height without leaving the ground.

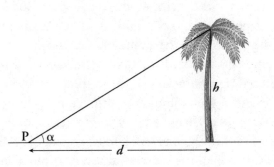

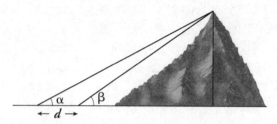

To measure the height of a mountain, we need to draw a picture that includes two triangles, illustrated above, as it is impossible to stand at the corner of the triangle directly underneath the summit. The surveyor solves this problem by observing the top of the mountain from two different points, which are in line with the summit, and which produce angles α and β. He or she also measures the distance d between the two points. The height of the mountain can be calculated from the values of tan α, tan β and d (I show how in Appendix Three on p. 295).

Triangle ratios transformed fields like navigation and warcraft, since they allowed sailors and soldiers to measure distances to objects they couldn't reach without either drowning or getting shot. The ratios also enabled Arab scholar Al-Biruni to obliterate Eratosthenes's record for the circumference of the Earth. In the eleventh century CE, Al-Biruni was staying at a fort in the Punjab's Salt Range when he came across the perfect geographical features to measure the height of a mountain: a high peak facing a flat plain. It would have been churlish not to measure the summit with trigonometry, so he did. Yet rather than pack his bags afterwards and move on, Al-Biruni instead climbed to the summit and measured the angle between his horizontal view and the horizon, marked θ in the illustration opposite. He joined both the point on the horizon where it met the Earth and the point on the summit where he stood to the centre of the Earth to make a right-angled triangle, from which he used trigonometry to deduce that the radius of the Earth was the height of the mountain multiplied by $\frac{\cos \theta}{1 - \cos \theta}$. (That proof is also in Appendix Three on p. 295). His calculation gave an Earth radius of 6335km, which gives a circumference of 39,800km, only 0.5 per cent less than the correct value, and almost ten times more accurate than Eratosthenes's estimate.

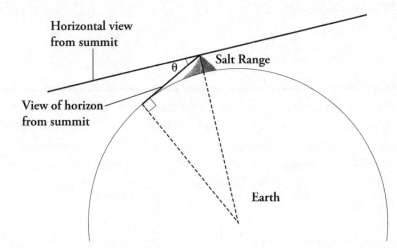

Al-Biruni measures the radius of the Earth.

Triangle ratios were game-changers for architects, astronomers, artillerymen, scientists and seafarers. Yet they also led to new abstract mathematics, throwing a novel perspective on geometrical classics like Pythagoras's Theorem. Refresh yourself with the theorem from the illustration on p. 62. It states that:

$a^2 + b^2 = c^2$, where c is the hypotenuse and a and b are the other two sides

If we let α be the angle between b and c, then:

$$\sin \alpha = \frac{a}{c} \qquad \cos \alpha = \frac{b}{c}$$

In other words $a = c \sin \alpha$, and $b = c \cos \alpha$, which we can substitute in the Pythagoras equation:

$$(c \sin \alpha)^2 + (c \cos \alpha)^2 = c^2$$

Which expands to:

$$c^2 (\sin \alpha)^2 + c^2 (\cos \alpha)^2 = c^2$$

Which reduces to:

$$(\sin \alpha)^2 + (\cos \alpha)^2 = 1$$

Nice! Here we have a concise formula that shows us how to calculate the sine from the cosine, or vice versa, without even needing to think about drawing a triangle. It is the simplest of what are called the 'trigonometric identities', equations that involve combinations of the trigonometric ratios. Arab mathematician Ibn-Yunus, a contemporary of Al-Biruni, is credited with introducing the following:

$$\cos \alpha \times \cos \beta = \frac{\cos (\alpha + \beta) + \cos (\alpha - \beta)}{2}$$

This formula was a showstopper, although it took mathematicians five hundred years to figure out why. The equation provides a way to turn multiplication, a difficult arithmetical operation, into addition, a much simpler one.

Imagine we want to multiply 0.2897 by 0.3165.

Both these numbers are between 0 and 1, so there are angles that have these numbers as cosines. We consult our trigonometric tables to find them. They are:

$$\cos 73.160° = 0.2897$$
$$\cos 71.548° = 0.3165$$

So we can write:

$$0.2897 \times 0.3165 = \cos 73.160° \times \cos 71.548°$$

The identity above tells us that this multiplication equals:

$$\frac{\cos (73.160 + 71.548)° + \cos (73.160 - 71.548)°}{2}$$

$$= \frac{\cos 144.708° + \cos 1.612°}{2}$$

Consulting the tables:

$$\frac{\cos 144.708° + \cos 1.612°}{2} = \frac{-0.8162 + 0.9996}{2}$$

$$= \frac{0.1834}{2}$$

$$= 0.0917$$

Which is the answer to our desired multiplication, 0.2897×0.3165. The answer is extremely accurate. Put the two numbers into your calculator and you will also get 0.0917, when rounded to four decimal places.

The procedure above might seem a complicated way to perform multiplication, but in the late sixteenth century it was by far the easiest method. Rather than write out a long multiplication, which is effortful and time-consuming, we just need to look in a book of tables, add two numbers, subtract two numbers, look in the book again, add two numbers and divide by two. The method is called 'prosthaphaeresis', from the Greek for addition and subtraction, *prosthesis* and *aphaeresis*.

Inspired by prosthaphaeresis, the Scotsman John Napier looked for an even better method to transform multiplication into addition. It led to his discovery, in 1614, of the logarithm. Instead of multiplying two numbers, one could add their logarithms. Napier's logs simplified the process of multiplication even further, and prosthaphaeresis became obsolete. Yet for a few decades of glory, the right-angled triangle – the quintessence of geometry – had a dual role as the hidden weapon of arithmetic.

Triangles are useful individually, but they are especially dynamic team players. If you draw a network of triangles, as illustrated overleaf, and measure all the angles in it, then you need only to measure one line in the network to be able to calculate the lengths of all the others. Suppose the only line whose length we know is the bold one, and it has length l. A trigonometric identity called the sine rule provides us with a formula for the other two lengths: $\frac{l \sin \beta}{\sin \alpha}$ and $\frac{l \sin \gamma}{\sin \alpha}$, where α is the angle opposite the bold line, and β and γ are

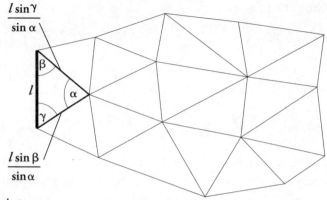

Triangulation.

the other angles. Since all the angles in the network are known, each new length can be used to calculate two new lengths, and so on until every length in the network is known. The method works for all triangles, not just right-angled ones.

In 1533 the Dutch mathematician Gemma Frisius realized that this technique, triangulation, was tailor-made for cartography, since angles are much easier to measure than large distances across land. His idea was to choose prominent points on a landscape, such that from each point you can see at least two others, and to turn these points into a network of triangles. He measured the angles between the points using a theodolite, which is essentially a circular protractor on a stand. By measuring a single distance in the network, the base line, he could deduce all the other distances using trig tables, and hence draw an accurate map.

France was the first country to undertake a national triangulation, which it did in 1668. The most difficult single challenge in any triangulation is to measure the baseline. Abbé Jean Picard used an 11km stretch of straight road on the outskirts of Paris, between a mill at Villejuif and the pavilion at Juvigny, which he measured meticulously with wooden rods. He then headed northwards, using landmarks like clock towers and the tops of hills as the points of his triangles, measuring only the angles between them. When Picard reached the Atlantic Ocean he discovered that the French coastline was significantly nearer to Paris than had previously been thought. 'Your work has cost me a major part of my realm!' harrumphed

King Louis XIV. Picard's triangulation continued for a century after his death, until France was covered by 400 triangles. The celebrated map of the nation it produced had more detail than any previously made, with almost the same scale as the standard yellow Michelin tourist maps available now.

The French had an *amour fou* for triangles. In 1735 Louis XV dispatched two teams of triangulators to the opposite ends of the Earth to settle a pressing scientific dispute. The Earth is not a perfect sphere. Whether it was flattened at the poles, like a grapefruit, or flattened at the equator, like a lemon, was hotly debated, and the subject of bickering between the British, who conjectured the former, and the French, who disagreed. The French realized that they could deduce the correct citrus shape by comparing the terrestrial distance covered by a degree of latitude near the North pole, and one near the equator. If the Earth was a perfect sphere, the distance covered by a degree of latitude would be the same everywhere – 360th of the distance round the world. But if the distance was longer near the pole, this meant that the globe flattened out there, and if the distance was shorter near the pole, the globe flattened out at the equator. The French sent one expedition to Lapland, and one to what is now Ecuador in South America. They used the stars to calculate the starting latitude, and in Lapland triangulated due north, and in Ecuador due south. At the end point of each triangulation, they again used the stars to determine the latitude. After battling snowstorms and mosquitos in Scandinavia, and altitude sickness in the Andes, the teams found that the length of a degree of latitude was longer in Lapland. The Brits were right in their choice of fruit: the world was indeed a giant *pamplemousse*.

The French used the triangle as a workhorse for their social and scientific advances. For Britain, on the other hand, it was an instrument of empire. The Great Trigonometrical Survey of India, which lasted for much of the nineteenth century, was the largest scientific undertaking of its time. In lives lost and money spent the cost is said to have exceeded many contemporary Indian wars. Starting at the southernmost tip, the Survey headed up through the jungle, the Deccan Plateau and the northern plains before reaching the Himalayas under the command of Colonel George Everest (pronounced EVE-rest).

During triangulations, both horizontal and vertical angles are measured. The network of triangles is three dimensional, allowing surveyors to calculate height as well as distance across. In the Himalayas, the altitudes of the peaks were of most interest. At that time, Chimborazo in Ecuador, measured by the French triangulation a century earlier, was regarded as the highest mountain in the world. The Himalayas were known to be snow-capped and majestic, but claims that they were higher than the Andes were dismissed as another tall tale from the land of snake charmers and rope tricks. This assessment changed when the survey reached a range of soaring peaks, the highest of which had no known local name. It was eventually named Mount Everest in honour of the colonel. It is the highest mountain in the world and one whose name remains forever mispronounced.

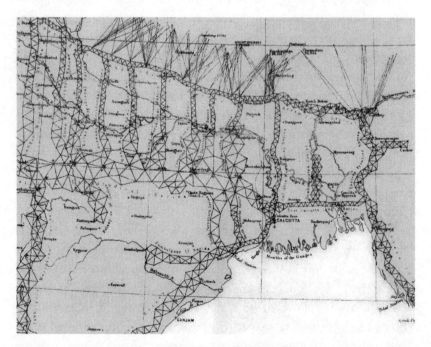

Northeast section of the Great Trigonometrical Survey of India, including Kolkata (formerly Calcutta) and the Himalayas.

Britain's first national triangulation took place between 1783 and 1853. (The site of one end of its baseline is now in a Heathrow Airport car park, where there is a small monument. Baselines and airports like flat land.) A 'retriangulation' was started in 1935 and lasted until 1962. The Ordnance Survey built more than six thousand concrete trig pillars on the points of the triangles, which provided the basis for a grid system still used in official maps.

The retriangulation, however, was out of date almost immediately. The need for national triangulations depended on angles being significantly easier to measure than distances across land. But in the 1960s a new technology, laser, enabled accurate measurement of long distances. Place a transmitter at one point and a receiver at the other, and a laser beam will travel between them at the speed of light. The distance from source to target is equal to the speed of light multiplied by the time elapsed. Once surveyors could use lasers they didn't need triangles any more.

Britain's 6200 remaining trig pillars are now sites of pilgrimage, not just to trig baggers like Rob Woodall, but to ramblers of all persuasions. The geometrical simplicity of the pillars – flat-topped, pyramidal obelisks – gives them an ageless mystical charm that, now they are battered and worn, makes you wonder if Druids placed them there rather than geographers.

New technology, however, still relies on triangles. Trigonometric ratios are an integral part of the Global Positioning System (GPS), the satellite-based infrastructure that pinpoints the locations of smartphones and in-car navigation computers wherever we are in the world. Each satellite in the network is on an independent orbit, determined by a set of parameters calculated using sines and cosines. For my phone to calculate its location, it must receive these parameters from at least four satellites. When it does, it processes the data, referring to a table of sines and cosines stored in its memory.

For two thousand years, scientists have used tables of trigonometric ratios. Nowadays we carry ours in our pockets. The principle that triangles with the same angles have sides in the same proportion to each other was used in the first proof in mathematics, and it remains essential in the information age.

Coneheads

Let's take a right-angled triangle, and upgrade it. Rotate the triangle around one of its short sides. The three-dimensional object produced is a cone, a solid with a circular base and a sharp tip at the summit. The shape is not practical: you can't roll it like a sphere or stack it like a cube, although historically the cone has found a niche in headgear. Vietnamese paddy farmers, wizards and dunces all don pointy hats. Indeed, the ancient Greeks wore the *pilos*, a conical hat made out of felt or leather. Their head for cones, however, was less sartorial than intellectual, for the cone is a mathematical treasure trove.

Slice through a cone with a knife and the outline produced is one of four distinct curves: the circle, the ellipse, the parabola or the hyperbola. The curves, known as the 'conic sections' are dependent

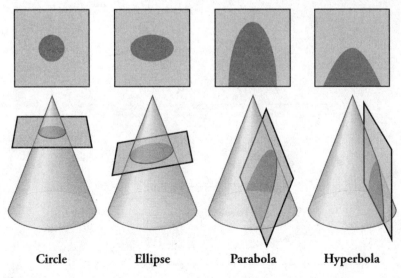

| Circle | Ellipse | Parabola | Hyperbola |

The conic sections.

on the angle of the blade. A horizontal cut makes a circle, a non-horizontal cut that still comes out the other side makes an ellipse, a cut parallel to the side makes a parabola, and all deeper cuts make a hyperbola, as illustrated on the previous page. Analysis of the conic sections was a high water mark of Greek geometry, and provides a spectacular example of a subject studied for its own sake that millennia later turned out to have momentous applications. The humble cone contained the answers to fundamental questions about the universe.

The circle is a familiar shape, the locus of points equidistant from a centre. Tie a pencil to a fixed pin and it will mark out a circle when the string is taut. Now place a loop of string between two pins, as illustrated below. The path of a pencil circumnavigating the pins while pushing the string taut is an ellipse. All circles have the same shape, meaning that if you take any circle you can shrink or expand it and it will be identical to any other circle. Ellipses, on the other hand, come in many shapes, depending on the relative position of the pins, or focus points. The closer the foci are to each other, the more an ellipse looks like a circle. When they meet, the ellipse *is* a circle. Mathematicians, in fact, relegate the circle to a special case of the ellipse, occurring when the foci coincide.

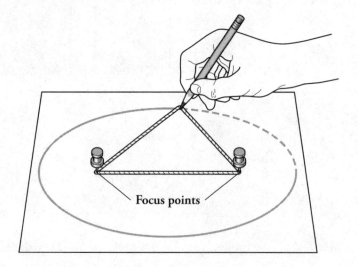

Focus points

Drawing an ellipse.

When we view a circle at an angle we see an ellipse. Wheels, coins, clocks, hoops, rings and discs always look like ellipses unless we observe them face on, which we rarely do. Likewise, for every ellipse, there is always an angle to view it from that makes it look like a circle. (Turn this book sideways and rotate it away from you to see the circle in any of the ellipses in these pages.)

The ellipse has a geometrical property of historic interest to the practitioners of indoor sports. If a pool table is made in the shape of an ellipse, a ball shot from one focus point will always rebound towards the other focus point – no matter in which direction the ball is shot. This rather cool feature is a consequence of the following property: the straight line from one focus to a point on the ellipse makes the same angle to the curve as the line from that point to the other focus, as shown below left. When you bounce a ball against the side of a pool table, the angle of the ball as it approaches the cushion equals the angle it makes when it leaves, as anyone who has ever chalked a cue tip knows, so the shot from one focus always rebounds in the direction of the other.

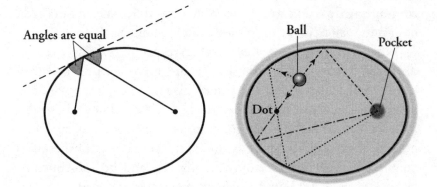

Lines from the curve to the foci make the same angle, giving pool players three indirect ways to pot a ball.

In the early 1960s, Art P. Frigo Jr., a high school student from Connecticut, built an elliptical pool table after studying the conic sections at his high school math club. Art's tables had a black dot at one focus point, a pocket at the second, and no pockets round the side. If there was just one ball on the table, as illustrated above right, there were three ways to pot it without aiming directly at the pocket: towards the black dot, in which case the ball would pass the dot, hit the cushion and rebound into the pocket; in the opposite direction to the black dot, in which case the ball would also rebound into the pocket; and in the opposite direction to the pocket, in which case the ball would rebound once, pass the black dot, and rebound again before reaching the target. The table was a pocketing machine! Art suggested that a game of 'Elliptipool' start with one white and six coloured balls on the table. The unorthodox shape opened up fascinating new patterns of play.

Art made his prototype at school and took it with him when he enrolled at Union College, Schenectady. The table was such a hit at his fraternity house that it made the TV news. He later patented it, and a national toy company offered him a deal. 'They had orders for 80,000 tables. I was 21 and I thought "I am going to become a millionaire!"' he said. The company hired Paul Newman, who had just starred in pool hall drama *The Hustler*, for the ad campaign. Yet there were hitches. The tables took almost a year to come to market, and when they did the wood warped easily. A sturdier, coin-operated version was made, which was rolled out to hundreds of bars in major cities. But it didn't catch on either.

When Art visited a venue to observe people playing, he was dismayed to see the table unused. 'It hurt me when I realized that people didn't understand the game,' he said. 'People perceived it as a round table just to be different. If you didn't know about the focus points the ball was not going in. People couldn't make a shot because they couldn't figure it out.' Yet, Art said, the experience taught him how *not* to launch a product, and he subsequently became a successful entrepreneur in diamonds and sponge mops. He now lives in Florida and imports olive oil.

The mathematical relationship between the foci of an ellipse may have failed to set American bar culture alight, but it has been put to dazzlingly good use by the lighting industry. Just as a billiard ball shot from one focus rebounds to the other, if a light source is placed at the focus of an ellipse made out of reflective material, all its beams will be reflected to the other focus. When you rotate an ellipse around the invisible line between the two foci you get a three-dimensional shape called an ellipsoid. Place a bulb at one of the foci of an ellipsoid shell with a mirror lining the inside surface and you have the main component of the theatre spotlight. There is no better way to make a sharp beam of light. The light emanating from the bulb is reflected through the second focus, providing a concentrated beam which is then refracted through a lens. The optical applications of the conic sections, in fact, explain the origin of the word 'focus': it is the Latin for 'fireplace'. In German its etymology is clearer – 'focus' is *Brennpunkt*, or burning point.

Buildings with ellipsoidal roofs have remarkable properties because sound made at one focus will rebound from everywhere on the roof towards the other focus. The giant dome of the Mormon Tabernacle in Salt Lake City, for example, was deliberately built in the shape of half an ellipsoid. If you drop a pin at the pulpit, which is situated at one focus, it can be heard clearly at the other focus 170ft away.

Ancient Greek mathematics lasted for almost a thousand years, from Thales, around 600 BCE, to Pappus, its last significant figure, around 300 CE. Sitting at the top table were three men: Euclid, Archimedes and Apollonius, the holy trinity of classical mathematicians. They all lived in the third century BCE. Euclid and Archimedes await us at a later stage. Apollonius, the youngest of the three, studied and taught in Alexandria, and he also lived in Pergamum in Turkey, home to the second-largest library in the Greek empire. Apollonius is now the least celebrated of the three Greek giants, although in his time he was known as *Megas Geometris,* The Great Geometer. Of his many books, only one survives: *Conics*, a treatise on cones.

In *Conics*, Apollonius showed how slicing through a cone produces the three types of section, and he introduced names for them. The ellipse comes from the Greek *leipein*, to leave out, the parabola from *para*, equal to, or beside, and the hyperbola from *hyper*, more than, or beyond. (The suffix *-bola* means 'thrown'.) The names Apollonius chose were based on the properties of the areas of these curves, which are too complicated to explain here, but we can understand what he meant by analogy to the slice angle, as illustrated on p. 79. When the angle of the slice is less than the angle or slope of the side, the section is an ellipse. When the slice angle is equal to the side angle, the section is a parabola, and when the angle is greater, the section is a hyperbola. *Conics* contains 387 propositions, and it is not an easy read, partly because Apollonius uses a cumbersome notation that is now obsolete. Still, his achievement was immense, and is regarded as the summit of Greek geometry. By so thoroughly investigating the properties of the cone, he laid the technical groundwork for major scientific advances two millennia later.

Apollonius loftily declared in *Conics* that here was a subject worthy of study for pleasure alone, yet he also worked on mathematics that had real applications. The earliest stargazers observed that planets do not move in straight lines – they meander across the skies, often temporarily looping backwards. (The word planet comes from the Greek *planetes*, wanderer.) Plato had declared that the planets move in perfect circles, the simplest, most elegant shape. This assertion was based on his certainty that the world is constructed with geometrical simplicity and elegance, even though the data suggested otherwise. His proclamation threw down the gauntlet for thinkers to explain the celestial meandering using some kind of combination of circular motions. Apollonius rose to the challenge, devising a system that became the standard model for almost two thousand years.

The Apollonian solution to planetary motion fixes the Earth at the centre of the cosmos. Each planet is rotating in a small circle, called an epicycle, which at the same time is rotating in a larger circle centred on the Earth, called a deferent, as illustrated opposite. The doily-like orbital path is just like the one drawn by a Spirograph, the toy in which a small cog with a pen rolls round a larger cog. The orbit of a planet rotating in an epicycle on a deferent has moments when it loops, which explains why planets occasionally appear to move backwards. Apollonius's system fitted the data with only minor errors, and the fit could be improved by introducing an extra epicycle into the arrangement. This meant that the planet was now subject to three circular motions: moving in a circle, which is going round a second circle, which in turn is orbiting a third circle around the Earth.

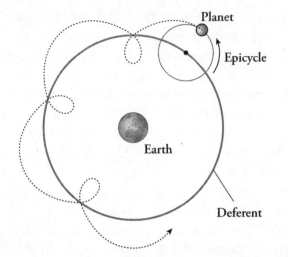

In the *Almagest,* written in the second century CE, the Greek astronomer Ptolemy described an epicycles-on-deferent system that was universally accepted as the true model for the cosmos until the sixteenth century. No one challenged the Ptolemaic programme, even when improvements in measurements required the addition of more and more epicycles to the model. In its final version, 39 cycles and epicycles explained the paths of the five planets, the Sun and the Moon. Plato's dream of geometric elegance had become a bewildering mess, criticized by even the Church as unintelligent design. 'If the Lord Almighty had consulted me before embarking upon the Creation, I should have recommended something simpler,' muttered the thirteenth-century king Alfonso X of Castile, also known as *El Sabio,* The Wise.

We now know that Apollonius was wrong. There *is* a simpler model for planetary orbits, which we'll come to shortly. Indeed, the derogatory phrase 'adding epicycles' is now used to describe bad science, the endless refining of a mistaken theory in the hope that eventually it will work. The epicycle system held sway for so long, however, because it did its job incredibly well. Usually a theory is overturned because it is shown to be false. The theory of epicycles, on the other hand, was never falsified because it cannot be. Remarkably, cycles and epicycles can be used to describe any closed and continuous orbit. Apollonius's idea was so good that no one needed to look anywhere else.

In 2005 the Argentinians Christián Carman and Ramiro Serra decided to describe an outrageously complicated orbit and then find the epicycles that produce it. They chose a picture of Homer Simpson because it looks nothing like an orbit and because – doh! – it's Homer Simpson. The mega-squiggle below is the model of the Homerian orbit. The large circle is the deferent, and the tangle of smaller circles contains 9999 epicycles of varying sizes. The planet is rotating around the 9999th epicycle, which is rotating around the 9998th, and so on, down to the first epicycle, which is rotating around the deferent. By the time the planet has completed one revolution of the deferent – when it will have completed two revolutions of the first epicycle, three of the second, and so on, including ten thousand revolutions of the 9999th epicycle – it will have traced the path of the cartoon figure. Carman and Serra were 'really excited and gratified' when they made it work. Plato, too, would have appreciated the poetry in Homer.

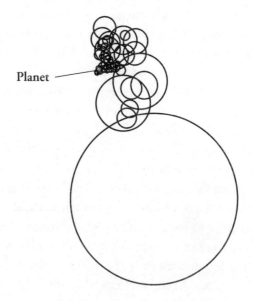

Looks like Marge, but it's Homer: the path made by a planet whose orbit is a combination of these 10,000 circles is a portrait of the Simpsons' paterfamilias.

At 4.37am on 16 May 1571 in the small German town of Weil der Stadt, Johannes Kepler was conceived. He was born 224 days, 9 hours and 53 minutes later, at 2.30pm on 27 December. We know these details from a horoscope Kepler cast for himself aged 26, in which he also reveals that he almost died of smallpox, his hands were badly crippled, he suffered continually from skin ailments, and when he lost his virginity aged 21 he did so 'with the greatest possible difficulty, experiencing the most acute pains of the bladder'. We can infer the traits that defined his life: hypochondria, introspection, an obsession with the stars and a love of numbers.

By the time he cast that horoscope, Kepler had published his first book, *The Mystery of the Cosmos*, in which he presented a model of the planetary system based on Nicolaus Copernicus's revolutionary theory, proposed half a century earlier, that the planets rotate around the Sun. Even though Copernicus had jettisoned geocentricism, however, he still believed the planets moved in epicycles. Kepler enhanced this view with a model in which the orbits of each planet fit into a superstructure of geometrical objects, the Platonic solids – which are the cube, the tetrahedron, the octahedron, the icosahedron and the dodecahedron – each of different sizes but with the Sun at the centre. The model was wrong, of course, but *Mystery* established Kepler as a name in learned circles, and when the celebrated Danish astronomer Tycho Brahe was building a new observatory near Prague, he appointed the ambitious young German as his assistant.

Brahe was a flamboyant aristocrat. He wore a prosthetic gold and silver nose, after a cousin sliced the original one off in a duel about a mathematical formula. He also had a pet elk, which fell to its death after drinking too much beer at a dinner. The Dane, however, was much more careful with his astronomical data, which was known across Europe as the most accurate and extensive ever compiled. He entrusted Kepler with trying to understand the orbit of Mars, the planet that wandered most from a circular path. The work was painful and laborious, involving the creation of possible orbits, the calculation of predicted positions, and checking with observed data: 'If this wearisome method has filled you with loathing,' Kepler later explained, 'it should more properly fill you with compassion for me, as I have gone through it at least seventy times.'

While Kepler was 'at war with Mars' he took a break to invent modern optics. *The Optical Part of Astronomy* includes a section on mirrors in the shape of the conic sections: the ellipse, parabola and hyperbola. Indeed, it was here that Kepler introduced the word 'focus' meaning the burning point of reflected light beams. On returning to Mars, exasperated by his inability to find a system of combined circular motions that agreed with the data, Kepler finally decided to ditch the theory of epicycles. He was hardly optimistic about this new course. 'I have cleared the Augean stables of astronomy of cycles and spirals,' he lamented, 'and left behind me only a single cartful of dung.' For a year Kepler experimented with an egg-shaped orbit, an oval with a flat end and a pointy end, even though he found such a shape repugnant, regarding it as neither symmetrical nor harmonic. To approximate the oval in his calculations, however, he used an ellipse – a geometrical prop familiar to him from his optical work on conic sections. And then it clicked – the prop could stand alone. '*O me ridiculum!* How ridiculous of me!' he exclaimed, 'There is no figure of a planet's orbit other than the perfect ellipse.'

Originally, Kepler had discarded the ellipse for the orbit of Mars because he thought it too simple an idea for previous scientists to have overlooked. Also, his knowledge that an ellipse has two foci seemed to contradict the Sun's uniqueness, which suggested it would have to be at the centre of the system, rather than at one

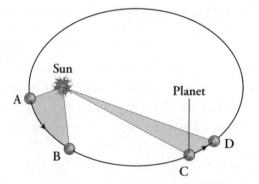

It takes the same time to get from A to B as it does from C to D, since the shaded segments have equal areas. A planet therefore moves more slowly when it is further from the Sun.

of two equally important points. He then realized that despite the apparent contradiction, the Sun *is* at one of the foci, and that its influence governs orbital speed. (There is nothing at the other focus.) A planet moving along an elliptical orbit moves faster the closer it is to the Sun, but spans out equal areas in equal times, as illustrated opposite. The philosopher Norwood Russell Hanson wrote that Kepler's breakthrough was the boldest exercise in imagination ever required in the history of science. 'Not even the conceptual upsets [of the twentieth century] required such a break with the past.' Apollonius's model of epicycles was ultimately supplanted by the ellipse, a curve The Great Geometer had named himself and whose properties he knew better than anyone else.

In 1610 Kepler received a message from Galileo Galilei, the eminent astronomer living over the Alps in Italy:

smaismrmilmepoetale v m i b u n e n u g t t a v i r a s

Galileo had news that was too exciting to keep to himself, but too valuable to tell anyone else lest it help their own inquiries. He announced it as an anagram, which established priority of discovery, kept the details secret, and avoided overcommitment just in case he was wrong.

The riddle drove Kepler crazy. Eventually, he believed he had solved it by rearranging the letters from gibberish into a statement that made sense: '*Salve umbistineum geminatum Martia proles*', or 'Hail, burning twin, offspring of Mars', although this involved a hopeful Latinization of the German word *umbeistehen*. Kepler was convinced his rival had discovered that Mars has two moons. Galileo later unscrambled the anagram as '*Altissimum planetam tergeminum observavi*', or 'I have observed the most distant planet to have a triple form'. The discovery was not about Mars at all, but about Saturn: the planet had a bulge on either side, both of which we now know are made by its rings. Remarkably, however, Kepler turned out to be right! Mars does have two moons, Phobos and Deimos, which were discovered more than two centuries later.

Galileo later teased Kepler with a second anagram, but this time the message made sense. It was deliberately provocative: '*Haec immatura a me iam frustra leguntur – oy*', or 'I am now bringing together these unripe things by me, oy!' Again Kepler found a meaningful solution: '*Macula rufa in Jove est gyratur mathem etc*', or 'There is a red spot in Jupiter which rotates mathematically' – and again he was wrong. What Galileo was, in fact, communicating was '*Cynthiae figuras aemulatur Mater Amorum*', or 'The mother of love [Venus] imitates the figures of Cynthia [the Moon]', meaning that Venus has phases, just like the Moon. Yet Kepler's misguided translation was again prophetic. Fifty years later, astronomers saw that Jupiter does indeed have a bloody splotch, a giant storm known as the Great Red Spot.

Galileo and Kepler changed the image of scientists from one of passive scholars to one of discovering heroes. With only a single universe out there, they wanted credit for mapping it first. Many others after Galileo, including Robert Hooke, Christiaan Huygens and Isaac Newton, employed the indecipherable anagram to protect their intellectual property, until in the eighteenth century publication in a journal became the standard method to announce the latest scientific advance.

Galileo accepted Copernicus's theory that the Earth orbited the Sun, yet he refuted Kepler's hypothesis that the planetary orbits were ellipses. Galileo did, however, make significant progress studying the motion of other types of spherical object. In the summer of 1592, as a young maths professor, he visited his friend and sponsor the Marquis Guidobaldo del Monte at his castle near Urbino. The Marquis's job as inspector of fortifications for Tuscany meant that he was especially interested in the trajectory of cannonballs. Do they fly off in a straight line, and then drop downward in a straight line, as conjectured by traditional Aristotelian mechanics, or do they follow some kind of curve before reaching their destination?

To find out, the two men conducted an experiment so simple it is hard to believe no one had done it before. They took a small metal ball covered in ink and launched it diagonally up an inclined plane. The outline left behind was a symmetrical arc. Galileo saw that objects go up exactly as they come down; the path of the rise mirrors the path of the fall. This symmetry led to the insight that motion can be separated into horizontal and vertical components. The side-to-side behaviour of an object in free flight is independent of the up–down behaviour. Galileo later conducted other experiments with inky balls and demonstrated that if a projectile is fired horizontally off a table:

(i) Horizontal displacement is proportional to the time elapsed. So, if a body travels 1 unit along in one second, it will travel 2 units along in two seconds, 3 units along in three seconds and so on.

(ii) Vertical displacement is proportional to the square of the time elapsed. So, if a body falls 1 unit in one second, it will fall 4 units in two seconds, 9 units in three seconds and so on.

From Galileo's knowledge of Apollonius and the conic sections, he deduced that the path of a ball fired off a table, illustrated below left, is a parabola. When a projectile, such as a basketball, is launched at an angle, shown below right, the path is also a parabola, but it must climb one leg before falling down the other. The parabola is the signature of a freely moving object under gravity – the jet of a fountain, the flight of an arrow or the hump of a ball kicked in the air. Novelist Thomas Pynchon named his magnum opus *Gravity's*

Ball fired horizontally

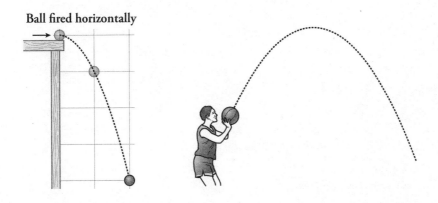

Rainbow after his description of the parabolic vapour arc left by a German V-2 rocket, a metaphor for how cultures rise and fall.

For almost two thousand years the conic sections were treated as the pinnacle of Greek mathematical thought, delightful curves but without practical function. Then, simultaneously, two applications were discovered hiding in plain sight. Planets travel in ellipses, and projectiles travel in parabolas. At the end of the seventeenth century, Isaac Newton demonstrated how both results followed from his laws of motion and universal gravitation. Galileo and Kepler had been studying the same problem on different scales. (Strictly speaking, a stone thrown in the air is actually beginning an elliptical orbit of the Earth, which it would complete if the Earth's mass was concentrated at its centre. However, from an observer's perspective, we can assume that stones are thrown in parabolas.)

An important and surprising fact about parabolas is that they all have the same shape. A parabola can be shrunk or enlarged so that it is identical to any other, just as a circle can be rescaled to be identical to any other circle. This follows from our original definition of the conic sections on p. 79, whereby each angle of slice produces a unique shape. For the circle and the parabola, only one angle is possible: horizontal for the circle, and parallel to the slope of the cone for the parabola. The ellipse and the hyperbola can be made from many angles, and therefore come in many different shapes.

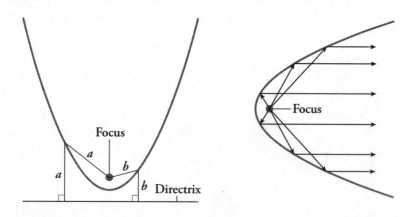

Geometry of the parabola.

The parabola can be defined in two other ways: it is the locus of all the points equidistant between a fixed point and a fixed line, which are known as the focus and the directrix, illustrated opposite on the left. And it is the curve that, when made out of a reflective material, reflects every beam of light emanating from the focus in parallel lines, as illustrated on the right.

The first definition provides origamists with an easy way to make a parabola. Mark a dot F on a sheet of paper, as shown below left. Take an arbitrary point P on the bottom edge and fold it on to the dot, as shown with the arrow, to produce the dotted fold line. Repeat for many arbitrary points on the bottom edge. The shape that emerges is a parabola. I leave it for the reader to work out why. (Tip: each fold represents the line of points equidistant between the focus and an arbitrary point.)

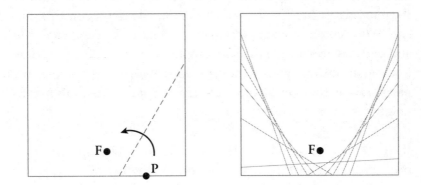

Folding a parabola.

The second definition is the reason why the parabola is the most common curve in a lighting shop. A bulb at the focus point of a parabolic mirror will reflect every beam in a parallel direction. Revolve a parabola around its central axis and you get a paraboloid, which is the shape of the reflective mirrors in torches, searchlights and car headlamps.

It works the other way round, too. Parallel rays of light entering a paraboloid will all be reflected towards the focus. So, if the purpose of a reflector is, say, to concentrate the sun's rays – which are effectively parallel since the sun is so far away – a parabolic shell is required. Solar energy technology relies on paraboloids. The Scheffler Reflector, for example, is a parabolic metal dish commonly used in the developing world for cooking. The dish is aimed at the sun and then rotates slowly to catch as much light as possible as the sun's position moves, while always reflecting the sunlight to the same spot, the focus, where there is a stove. The world's most powerful solar furnace is a 54m-high parabolic mirror at Odeillo in the French Pyrenees. Because of its size, the mirror is stationary and receives reflected sunlight from 63 smaller, rotating flat mirrors. At its focus is a circular target that on sunny days can reach temperatures of 3500°C, high enough to boil lead, melt tungsten, and incinerate wild boar.

Parabolic dishes are also used to reflect electromagnetic and sound waves arriving from distant objects to a focus. They are a familiar feature of the urban landscape, most commonly fixed to the roofs of satellite TV viewers' homes, but also found on air traffic control towers and military installations. Spies, TV sound engineers and birdwatchers use parabolic microphones to pick up quiet noises from many metres away. The principle is always the same. The paraboloid is the only shape that reflects parallel waves to a fixed point.

In 1668 Isaac Newton built the first 'reflecting' telescope, in which the fundamental components were mirrors, rather than the lenses that had been the key elements in previous telescopes. He realized that the best shape for the primary mirror was a paraboloid, although he was unable to manufacture one and instead made do with a spherical mirror. Even with this imperfection, however, the reflective telescope was much better than previous models and most telescopes made since the seventeenth century have been reflective ones.

Newton also made a discovery about parabolas that was originally only of theoretical interest, but which is now used by the telescope industry. When you rotate a cylindrical container of liquid, the surface takes the form of a paraboloid. The spinning causes the liquid to rise nearer the edges and to form a depression in the centre such that the cross section is a parabola. This property means that one way to make a parabolic mirror is to rotate molten glass in a container and let it set. The Large Binocular Telescope, one of the most powerful telescopes in the world, was made this way. It contains two 8.4m-diameter parabolic mirrors that were made in a giant rotating furnace in a basement lab underneath the University of Arizona football stadium in Tucson. Even though the lab can produce only one mirror a year, at a cost of tens of millions of dollars, the method is much cheaper and quicker than producing a similar mirror by grinding glass.

Cheaper still is the 'liquid mirror telescope', in which a telescope is constructed above a rotating drum filled with reflective liquid. The Large Zenith Telescope near Vancouver has a 6m-diameter circular pan filled with mercury, which becomes a paraboloid mirror when rotated. It is by far the most inexpensive of the world's large telescopes, but has one disadvantage: the pan rotates on the horizontal plane, so the telescope can only point directly upwards, to the zenith.

The solar oven at Odeillo, France.

In 1637 the French mathematician René Descartes invented coordinates, which led to the greatest advance in understanding the conic sections since the days of Apollonius. Cartesian coordinates determine position on a plane by reference to a vertical and a horizontal axis. Every point has a unique coordinate, (a, b), which is a along the horizontal and b along the vertical, as illustrated in (i) opposite. The importance of coordinates was that they enabled mathematicians to describe curves with equations, and to represent equations as curves. They therefore provided a bridge between geometry, the study of shapes, and algebra, the study of equations, which until then had been distinct mathematical fields.

Conventionally, we write equations using the variables x and y, which when plotted on a graph refer to the horizontal and vertical positions; in other words the coordinates (x, y). The graph of the equation $x = y$, for example, is the set of all the points (x, y) where $x = y$. It includes the points $(1, 1)$, $(2, 2)$ and $(3, 3)$, and is illustrated in (ii). The graph of $y = x^2$, on the other hand, is the set of all the points where $y = x^2$, and it includes the coordinates $(0, 0)$, $(1, 1)$, $(2, 4)$, $(3, 9)$, and so on. That curve, illustrated in (iii), is a parabola which touches the horizontal axis at the origin, or $(0, 0)$. In fact, since the school curriculum is more concerned with algebra than with geometry – the conic sections, for example, are not taught any more – our first encounter with the parabola usually comes when we plot the coordinates of $y = x^2$. You may recognize it as an old friend, the introductory U of elementary mathematics.

Algebra has its roots in solving practical problems. For example, what is the formula for the area of a square? If we let x be the side of the square, and y be its area, the formula is $y = x^2$. When an equation has an x^2 or a y^2 in it, but no higher power of x or y, it is known as a 'quadratic equation'. The Babylonians devised their own methods for solving quadratic equations, in particular for problems concerning the calculation of areas. By the Renaissance, quadratic equations were an extremely well-studied concept. What more could possibly be known about them?

Thanks to Cartesian coordinates, the quadratics were revealed to be the conic sections. In other words, every quadratic equa-

(i)

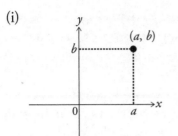

(ii)

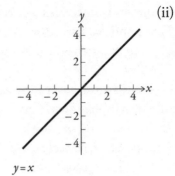

$y = x$

(iii)

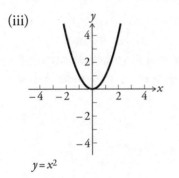

$y = x^2$

(iv)

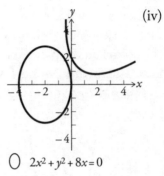

$2x^2 + y^2 + 8x = 0$

$16x^2 - 24xy + 9y^2 - 38x$
$- 84y + 121 = 0$

Cartesian coordinates.

tion describes a conic section, and every conic section can be described by a quadratic equation. Two of the most researched and pondered-over areas of mathematics were nothing but alternate representations of each other. The general quadratic equation $Ax^2 + Bxy + Cy^2 + Dx + Ey + F = 0$, where A, B, C, D, E and F are constants, and at least one of A, B and C is non-zero, always plots a conic section on a coordinate graph, and vice versa: any conic section drawn on a graph can be expressed by the above equation. In the illustration (iv) above, the equation for the ellipse is $2x^2 + y^2 + 8x = 0$, and the equation for the parabola, which sits diagonally on the page, is $16x^2 - 24xy + 9y^2 - 38x - 84y + 121 = 0$.

———

97

In the mid-nineteenth century the German mathematician August Ferdinand Möbius discovered a striking property of the parabola $y = x^2$: the curve is a *Multiplicationsmaschine*, or 'multiplication machine'.

Möbius had an eye for geometrical twists – literally, in the case of the Möbius Strip, his band of paper with a twist in it, and, more abstractly, in working out how the parabola could perform arithmetic. The technique is illustrated below left. To multiply $a \times b$, draw a straight line between the points on the parabola where $x = -a$ and $x = b$. The line crosses the y-axis at the answer! All that is required is an ability to draw a line and note the point of intersection. Below right is an example: 2×3. The line goes through the points on the parabola where x is -2 and 3, and crosses the y-axis at 6. The method works for any two digits (I have included a proof in Appendix Four on p. 297).

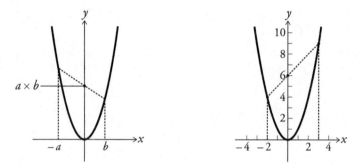

How to multiply two numbers with a parabola.

Möbius introduced his whimsical multiplication machine in 1841 in a footnote in the august *Journal für die reine und angewandte Mathematik*, the *Journal for Pure and Applied Mathematics*. He never mentioned it again, yet the idea of solving arithmetical problems using geometry was later reinvented by a young French engineering student, Maurice d'Ocagne. He discovered that many sums other than $a \times b$ can be calculated by drawing a straight line between two points on a graph and reading off the answer. In 1891 d'Ocagne coined the term 'nomogram' for any chart that could be used for computation in this way, and he designed lots of them. Each nomogram works for one formula only. The formula in the illustration below, which describes the velocity of water flowing through a rectangular opening in a weir, where V is the velocity and h_1 and h_2 are the heights of the upper and lower edges of the opening, is solved by the accompanying nomogram, from 1921. A straight line through the values for h_1 and h_2 on the curved line will hit the correct value for V on the vertical line. All you need to solve the cumbersome equation is a ruler and a steady hand. Nomograms eliminated laborious and time-consuming calculations. They were used widely in engineering and the military until the 1970s, when the electronic calculator made them instantly obsolete. Ingenious, practical – and often beautiful – nomography is now a forgotten art.

———

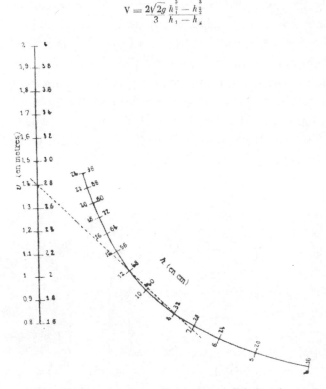

Calculating aids called 'nomograms' were in common use before the invention of the pocket calculator. This one from 1921 concerns the flow of water in a weir.

The hyperbola stands out from the other conic sections because it has two parts. To understand why, we need to return to our original definitions of slices through a cone (see p. 79). The full picture would reveal that our knife is actually cutting through a double-cone, which is a cone standing upside down on another one identical to it. For the ellipse and parabola the cutting angle means the slice never reaches the top cone. On the other hand, the angles that produce hyperbolas always slice through both top and bottom, as illustrated below in (i), producing a pair of symmetrical U-shaped wings.

The hyperbola introduced a beguiling new concept to geometry: the 'asymptote', another term coined by Apollonius. An asymptote is a line that is approached infinitely closely by a curve but never touched. The hyperbola is framed by two intersecting asymptotes, shown in (ii). Each open section of the curve is constantly moving towards the asymptote, but at no point will curve and line meet. 'I believe that if the geometrician were to be conscious of this hopeless and desperate striving of the hyperbola to unite with its asymptotes,' wrote the Spanish thinker Miguel de Unamuno, 'he would represent the hyperbola to us as a living being and a tragic one!' The hyperbola is often seen in the home. It is the shape of the arched corrugations of a sharpened pencil (the tip is the cone; the flat side is the slice), and the shadow made by a lamp (the beam is the cone; the wall is the slice), illustrated in (iii) and (iv).

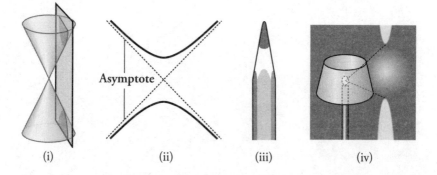

Hyperbolas.

The hyperbola has two foci, just like an ellipse. A good way to think of the hyperbola is as an ellipse that goes to infinity in one direction and then comes back the other way. We can also define the hyperbola by the properties of these foci, as we did the ellipse. The hyperbola is the path of a point whose distances between two foci have a constant difference, whereas for the ellipse they have a constant sum. In the illustration below left, a is the distance from an arbitrary point P to one focus, and b is the distance from P to the other focus. The hyperbola is the locus of P when $(a - b)$ is a fixed value. By considering $(b - a)$ we get the other wing. We can also define the hyperbola by the behaviour of light. Beams from a source at one focus will be reflected outwards from a hyperbolic mirror in an opposite direction to the other focus, illustrated below right. Ritchey–Chrétien telescopes, the most common type of large astronomical telescope, contain hyperbolic mirrors.

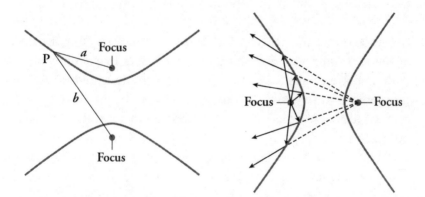

Geometry of the hyperbola.

I have provided methods to make an ellipse and a parabola, so I feel duty bound to extend that pleasure to the hyperbola. This time we will construct a model in three dimensions. We will make a hyperboloid, a shape that looks like a 1970s plastic stool, which is what you get when you rotate a hyperbola around its axis, illustrated overleaf on the left. For its construction we need two circles made of card and several pieces of string. The first step, illustrated in the

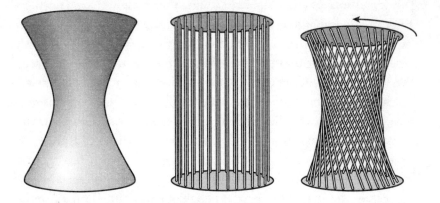

A hyperboloid, and how to make one with string.

centre, is to thread the string from one circle to the other to create the outline of a cylinder. The second step, illustrated on the right, is to rotate one of the circles. The shape that appears is a hyperboloid.

In the seventeenth century Christopher Wren, a young English astronomy professor, saw a round wicker basket in a shop window with a pattern just like the string model illustrated above. It prompted him to realize a stunning property of the hyperboloid: that even though the surface is smooth and curved, it consists uniquely of straight lines. Wren saw immediately how this fact could be exploited to make hyperboloids out of solid material using a single straight blade. Imagine you have a cylindrical piece of clay on a potter's wheel. Place a blade at a diagonal to the cylinder so that it indents slightly into the clay. If you keep the blade in the same position when you spin the wheel, after one rotation the clay cylinder will be a hyperboloid. Wren was interested in making hyperboloid lenses for telescopes. He did not foresee that centuries later his discovery of the straight-line property of hyperboloids would be dramatically employed in architecture, the field in which he subsequently became far better known.

In the nineteenth century the French maths educator Théodore Olivier designed several models of hyperboloids and related three-dimensional conic shapes to use as teaching aids. Made with wood and metal frames and coloured strings, they were much in demand by universities. Several ended up on display at London's Science Museum, where in the 1930s they so entranced the British artist Henry Moore that he started to use string in his sculpture. 'It wasn't the scientific study of these models but the ability to look through the strings as with a bird cage and see one form within the other which excited me,' he said. Olivier's string models are beautiful objects that mesmerize like an optical illusion, presenting curved surfaces that on closer inspection are made up of straight lines. (Olivier's personal set of models was acquired in the late nineteenth century by Union College Schenectady, where years later Art Frigo launched Elliptipool.)

The top circle in the string model opposite is rotated clockwise, so the strings at the front slope like a \ . If the circle had been rotated an equal angle in the other direction, an identical hyperboloid would have been produced, in which the front strings slope like a /. In order for a hyperboloid wicker basket to be robust it

Keeping cool with hyperboloids.

will be made of canes that slope in both directions. On a larger scale, a hyperboloid made from a lattice of steel girders is unusually sturdy, which provides a technique for building very large curved structures using only straight beams. The first hyperboloid in architecture was a 37m-high water tower built in Nizhny Novgorod, Russia, in 1896, and many more have followed. Concrete cooling towers at power stations are hyperboloids, as is the 600m-high Canton Tower in Guangzhou, the fourth tallest freestanding structure in the world.

I introduced the hyperbola last, but it is the one conic section that we have seen before. When two properties are inversely proportional to each other, like word frequency and rank in James Joyce's *Ulysses*, their mathematical relationship can be expressed as $y = \frac{k}{x}$, where k is a constant. This equation describes a hyperbola, where the asymptotes are the horizontal and vertical axes. Many natural laws involve properties that are inversely proportional to each other – like Boyle's law, for example, which states that the pressure of a gas is inversely proportional to its volume – so hyperbolas are ubiquitous in science. The common statistical term 'long tail' is often a euphemism for a hyperbola and its asymptote.

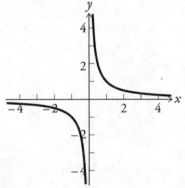

The curve $y = \frac{1}{x}$ is a hyperbola.

We started this chapter by defining the conic sections as exactly that – the slices of a cone – and then looked at the properties of each one individually. We'll finish with a final, all-inclusive definition: the conic sections are the curves whose distances between a point (the focus) and a line (the directrix) have a constant ratio. When the ratio $\frac{\text{curve to point}}{\text{curve to line}}$ is bigger than 1, meaning that the curve is always proportionally closer to the directrix than the focus, we have a hyperbola, as illustrated below. When the ratio is 1 we have a parabola, and when it is less than 1 we have an ellipse. These ratios are better known as the 'eccentricities' of each curve, since they are a measure of how much the curves deviate from the circle. In the illustration, the three curves share the same focus, F, and the same directrix. The ellipse has eccentricity 0.75 and the hyperbola 1.25.

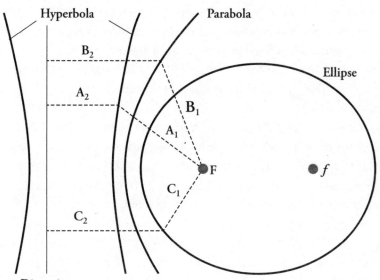

Conic section	Hyperbola	Parabola	Ellipse	Circle
Eccentricity	$\dfrac{A_1}{A_2} = k > 1$	$\dfrac{B_1}{B_2} = 1$	$\dfrac{C_1}{C_2} = k < 1$	0

The conic sections: a family of eccentrics.

Now imagine that you are an astronomer, and that the illustration is a model of the solar system. Let F be the Sun. The conic sections with focus F are the set of all possible celestial orbits.

Planets orbit in ellipses: the Earth's orbit has eccentricity 0.0167, very close to a circle. The faster an object travels along its orbit, the higher its eccentricity. Halley's Comet, for example, has twice the orbital velocity of the Earth. Its orbit looks like a surfboard with the Sun at the nose, which is why for most of its 75-year orbit it is too distant to see with the naked eye. The comet's eccentricity is 0.967, close to becoming a parabola. When a comet's eccentricity *is* 1, its orbit is a parabola, which means that it will pass the Sun only once in its lifetime before heading off for ever. When a comet's eccentricity is more than 1, its orbit is a hyperbola. This type of comet is rare, however, and the ones that have been spotted have velocities only fractionally higher than what is required to escape from the ellipse. The comet C/1980 E1, spotted in 1980, had an eccentricity of 1.057, the highest ever recorded.

Imagine that the directrix and the focus F in the diagram are fixed. We're going to see what happens to the conic sections as we vary the eccentricity. When the eccentricity is zero, the curve is a dot at F. Now slowly increase the eccentricity from 0 to 1. An ellipse appears, and it gets bigger and bigger. Since F is fixed, the other focus, *f*, will be moving rightwards as the ellipse grows. When the eccentricity is 1, *ping!* the ellipse becomes a parabola and *f* is at infinity. When you increase the eccentricity from 1, the curve becomes a hyperbola and *f*, the second focus, appears on the left-hand side of the page. As the eccentricity rises, the curves are all hyperbolas and *f* continues moving rightwards. Johannes Kepler, in *The Optical Part of Astronomy*, was the first person to suggest that the conic sections could morph into each other like this. Like many of his other ideas, it was a watershed: a fresh insight into two concepts that philosophers had struggled with until then – continuity and infinity. It was an important step towards a new way of doing maths. We will return to the great German and his treatment of these ideas at a later stage, in our discussion of the infinitesimal calculus.

———

The conic sections are one of the great legacies of Greek mathematics: simple to describe, observable everywhere, the subject of beautiful theories and timeless applications. I may have given the impression that the circle is the least interesting type of ellipse. Far from it. The circle deserves a chapter of its own.

Bring on the Revolution

The circle is the embodiment of geometrical perfection: smooth everywhere, harmonious and symmetrical. It is the locus of a point equidistant from a centre, the simplest two-dimensional shape there is. But when we divide the distance *around* the circle (the circumference) by the distance *across* it (the diameter), we hear a scream:

> 3.14159265358979323846264338327950288419716939937510582097494459230781640 6286208...

This number, the numerical value of the circumference divided by the diameter, is constant for all circles and its decimal digits continue for ever with untamed insubordination. In the eighteenth century the number gained its own name and symbol, pi, or π, and it has become a cross-cultural icon, the most famous constant in science and a metaphor for the inscrutability of the universe. Everyone learns it at school and for many it is the only thing they remember about mathematics.

Now here's the thing.

Pi is wrong.

The calculation is correct, of course. The ratio $\frac{\text{circumference}}{\text{diameter}}$ is evidently the above number beginning 3.14. Pi is wrong because it is an inappropriate number to represent the circle. Pi is an imposter, a false idol undeserving of its international acclaim.

Or so wrote Bob Palais, an American mathematician, in 2001. He argued that a much better choice for the circle constant would be the ratio $\frac{\text{circumference}}{\text{radius}}$, because the radius, the distance from the centre to the side, is a much more fundamental concept than the diameter. Many agree with him, including me. Look at our definition of the circle in the opening paragraph. The circle is a fixed distance, the radius, rotated around a centre. The diameter, or width, is an afterthought. Mathematics is a quest for elegance, clarity and correctness, and it is unfortunate that the most famous number in maths does not reflect the truth about circles in the clearest, most elegant and most correct way. (At school we are taught the word 'diameter' purely to understand pi, and once we have learned what it means we never return to it. Mathematicians always take for granted that the diameter is just the radius times two.)

In 2010 Michael Hartl, a Silicon Valley entrepreneur, ratcheted up anti-pi sentiment by baptizing the ratio $\frac{circumference}{radius}$ with the Greek letter tau, or τ. Tau is twice pi because the circumference fits twice as many radiuses around it as it does diameters. In other words:

$$\tau = 2\pi = 6.28318530717958647692528 6766\ldots$$

Just like pi, tau's decimal expansion is infinite and follows no known pattern.

In the Tau Manifesto, Hartl encourages young mathematicians to replace pi with tau in their work. A start would be to preface all papers with 'For convenience, we set $\tau = 2\pi$'. He warns that the fight will be long because the foe is so powerful, the beneficiary of centuries of propaganda. 'Some conventions, though unfortunate, are effectively irreversible,' he writes. '[But] the switch from π to τ can ... happen incrementally; unlike a redefinition, it need not happen all at once.'

The symbol τ is triply clever. It looks like a one-legged π, so if we see these symbols as fractions where the number of legs is the denominator – that's the number under the line – then τ really is twice π, since a quantity divided by 1 is twice something divided by 2. Tau also works as an abbreviation for 'turn', in the same way that pi was originally an abbreviation for 'periphery'. And just as pi is a delicious homonym of pie, a dish most often presented in circular form, so too is tau of Tao, the Chinese spiritual path whose symbol ☯, the yin and yang, expresses harmony and movement within the circle.

The Tau Manifesto makes a serious point in a light-hearted way. The quintessence of a circle is the turn of its radius, not the span of its width. Indeed, the dynamic properties of a circle, as exemplified by the wheel, are the mechanical ground rules on which civilization is based. In this chapter you will learn that the three most important things about circles are rotation, rotation, rotation.

Let's roll.

The path of a point on a rolling wheel may be unlike any curve you have seen before. It certainly was for Galileo, who named it the 'cycloid' and was the first to study it in any depth. The father of modern mechanics, Galileo was naturally drawn to curves that originate mechanically. The wheel moves smoothly, yet creates a curve with sharp spikes, the cusps, where its direction reverses. Each hump corresponds to one full rotation, the completion of a cycle. The cycloid looks less like a curve than a queue of sleeping tortoises.

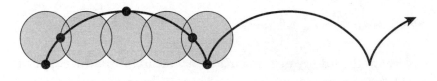

The cycloid.

In the illustration I have marked the position of the point at each quarter-turn, and you can see quite clearly that it covers more distance when it's in the upper half of the wheel than when it's in the lower half. The wheel moves in two ways: horizontally along the ground and rotationally around its centre, and both motions combine differently throughout the cycle. If the wheel is rotating at constant speed, the point on the wheel reaches its highest speed with respect to the ground at the crest of the cycloid, and its lowest at the cusp, where it reaches zero instantaneously before accelerating again. It is startling to realize that for any rolling wheel – even one on a racing car doing 200mph – the point in contact with the ground is stationary. Painters know that the top half of rolling wheels go faster than the bottom halves, which is why they often draw the top half of a moving wheel blurred and the bottom half in focus. Likewise, the spokes of a rolling bicycle wheel are often only visible near to the ground, where they are moving slowly enough to be seen.

A train wheel consists of two discs, one that sits on the track, and the rim, or flange, which dips over the side. A point on the flange will trace a curve that curls back on itself when below the level of the track, as shown in the illustration. On all trains, therefore, there are always points on the wheel that are going against the direction of travel.

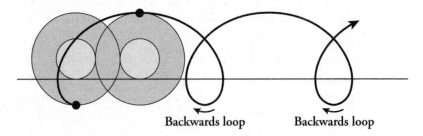

Backwards loop   Backwards loop

The path of a point on a train wheel.

Never in the history of mathematics has a curve been the object of such frenzied attention as the cycloid was in the seventeenth century. So beautiful its shape and so bitter the squabbles among its admirers that it gained a reputation as the 'Helen of geometers'. Suitor *numero uno*, Galileo, was practical in his advances. He cut the shape of the curve from a piece of material and judged it to be pi times heavier than a piece of the same material cut in the shape of the generating circle. He concluded that the area under the curve was therefore pi times the area of the circle. He was close – but wrong. The area is exactly three times larger, a result proved later by the French mathematician Gilles Personne de Roberval.

Roberval (1602–75) proved many theorems about the cycloid but did not publish any of them. To keep his position as professor of mathematics at the Collège de France, the country's most prestigious seat of learning, he had to provide the best answer to a problem announced publicly every three years. Roberval, as professor, also set the problems, which meant he had no incentive to share his results lest they help potential rivals with an eye on his job. His position gave him prestige and a salary, but castrated his legacy. He is probably the least remembered of the great French mathematicians. He was also famously irascible, and became bitterly frustrated when others announced results he claimed to have proved already. When his friend, the Italian Evangelista Torricelli, published the first treatise on the cycloid in 1644, a fuming Roberval dispatched a letter accusing him of plagiarism. Torricelli died three years later, of typhoid, but a rumour circulated that his death was due to the shame of being accused of such dishonour.

One evening in Paris in 1658, Blaise Pascal lay awake in bed tormented by a vicious toothache. Once a celebrated mathematician, Pascal had abandoned maths to concentrate on theology and philosophy. To take his mind off the pain, he decided to think about the cycloid. Miraculously, the toothache disappeared. Surely, he thought, it was a nudge of encouragement from God that he should continue to investigate this divine curve. He thought about the cycloid for eight industrious days, discovering many new theorems. Rather than publishing them, however, he turned them into an international challenge. He invited colleagues to prove some of his results, offering forty Spanish gold pieces for first prize and twenty for second. Only two mathematicians answered his call, John Wallis in England and Antoine de Laloubère in France. But their submissions contained errors, so Pascal did not award the prize, enraging both men. Instead he published his results in a pamphlet. He also received a letter from Christopher Wren, which demonstrated a fact of which the Frenchman was unaware. Wren had answered possibly the most basic question you can ask about the cycloid: how long is it? Wren showed that it is exactly eight times the radius of the generating circle. When Roberval found out, of course, he was incensed, insisting he had proved this assertion years before.

The cycloid's allure grew even more when Christiaan Huygens discovered a remarkable mechanical property it possessed. The Dutch scientist was experimenting with pendulums as a new basis for the design of clocks. A simple pendulum is a piece of string with a bob at the end of it, as illustrated below left. The bob's path is a circular section and the further it swings from the vertical position the longer it takes to complete a full swing. To use a pendulum for timekeeping, however, Huygens wanted the bob to complete each swing in an equal time interval regardless of how wide the swing was. Inspired by his friend Pascal's challenge, he realized that the path he was looking for was none other than an upside-down cycloid, illustrated below right, and that this path could be achieved by placing two cycloid 'cheeks' at the top of the pendulum. When the pendulum swings, its string bends around each cheek, pulling the bob from its circular path into a cycloidal one. However far from the centre the bob swings on a cycloidal pendulum, it takes exactly the same amount of time to return to its starting point.

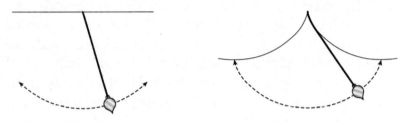

A simple pendulum, and one swinging between two cycloids.

Another way to marvel at this property is to imagine balls sliding down a frictionless track which is in the shape of an upside-down cycloid, as shown below. All balls take exactly the same time to reach the bottom, irrespective of their starting positions. A ball placed higher up starts off on a steeper slope than a ball starting at

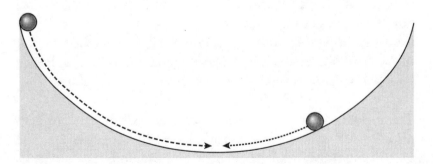

The path of equal descent.

a position on the curve below it, giving it greater initial acceleration, and therefore a higher speed. The two balls will collide precisely at the bottom of the curve. When the cycloid was heralded as 'the path of equal descent' – or the *tautochrone*, from the Greek *tauto*, meaning 'same', and *chronos*, meaning 'time' – scientists were smitten all over again.

The apogee of the curve's narrative arc came as the seventeenth century drew to a close. An article in the *Acta Eruditorum*, a new scientific periodical printed in Leipzig, trumpeted:

> *I, Johann Bernoulli, address the most brilliant mathematicians in the world. Nothing is more attractive to intelligent people than an honest, challenging problem, whose possible solution will bestow fame and remain as a lasting monument ... If someone communicates to me the solution of the proposed problem, I shall publicly declare him worthy of praise.*

The problem of which he spoke – and to which he already knew the answer – was this question: *What is the path of quickest descent?* Or, what is the shape that a frictionless slide must be in order to get an object from one point to another in the quickest time? The desired curve was nicknamed the *brachistochrone*, from the Greek *brachistos*, meaning 'shortest', and *chronos*, meaning 'time'. Bernoulli said that the curve was not a straight line, but was nevertheless well-known. Hint! The answer, if you haven't guessed already, was the cycloid. The illustration below shows the fastest

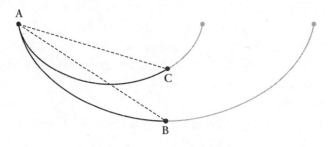

The path of fastest descent.

descent between the points A and B, and between A and C. Since the cycloid has only one shape, the curve must be scaled to fit, depending on the relative positions of the start and end points. The path may only drop, as it does between A and B, or it may drop and then rise, as it does between A and C. When the path drops and rises, the advantages of a steeper and longer drop at the start outweigh the slowing effect of the rising curve at the end. If you make a model of an upturned cycloid, and roll a metal ball down it from, say, point A to point B while simultaneously rolling a ball down a straight slide (the dashed line) from A to B, the effect is striking, even though you know which ball will win. Compared to the speedy ball descending the cycloid, the ball descending the straight edge looks like it is rolling through mud. From the eighteenth century onwards, universities and museums started building wooden cycloids to demonstrate the *brachistochrone*. These models could equally demonstrate the *tautochrone*: place a ball on each side of the upturned cycloid and, regardless of where they start, the balls will knock into each other at the bottom-most point.

After six months Bernoulli had received only one correct response to his challenge, from his German friend Gottfried Leibniz. So he published a new appeal for entries in the *Acta*, noting the absence of 'those who boast that through special methods … they have not only penetrated the deepest secrets of geometry but also extended its boundaries in marvellous fashion'. The remark was a jibe at Isaac Newton and his method of fluxions, a powerful new mathematical tool that promised to solve problems like the *brachistochrone*, and which we will cover in a later chapter. Bernoulli sent Newton a copy of the *Acta* to make sure he got the message. Newton was in his fifties at the time, no longer a Cambridge professor but instead in charge of the Royal Mint at the Tower of London. He read Bernoulli's letter after arriving home tired from work, and did not go to sleep until he had nailed the answer, at four o'clock in the morning. 'I do not love … to be … teezed by foreigners about Mathematical things,' he grumbled. Newton submitted his entry anonymously. Upon reading it Bernoulli is said to have declared '*ex ungue leonem*' – 'I recognize the lion by his claw.'

And so the cycloid, already the subject of so many quarrels, became the flashpoint of an early skirmish in the greatest feud in Enlightenment science. Newton's fluxions were mathematically equivalent to Leibniz's infinitesimal calculus, and, as we will see, a nasty priority dispute developed between both men that set the English and European scientific communities against each other for a century. The egos of men, however, did nothing to tarnish the sanctity of the curve. On the title page of Bernoulli's collected works is a picture of a dog gazing lovingly at a cycloid, together with the motto *Supra invidiam* – above envy.

Since the cycloid is the path of quickest descent, you might think it would be the favoured shape for a skateboard ramp. Yet as far as I can discover only one exists, built by the French artist Raphaël Zarka in New York in 2011 as part of a project mixing physics, sculpture and public spaces. The skaters did not like it, since it felt so unusual. 'If I were a completely round ball-bearing dropped from the top of the cycloid ramp, I would probably be able to better gauge the true ascent and descent,' said skate writer Ted Barrow, 'but as a skateboarder who has struggled with mastering a set of skills that all come down to attempting to maintain balance and NOT fall off his skateboard when speed increases, my experience was more about adjusting the speed, and suiting my movements to the strange curves of the walls, than experiencing the fastest descent.' The cycloid skateboard ramp, he added, is unlikely to catch on.

The cycloid belongs to a family of curves called the 'roulettes', each made by the path of a point on a rolling wheel. Roulettes roll on all terrains. The path of a wheel rolling around a circle of identical size is a cardioid, illustrated below left, so-called because it looks like a heart. The path of a wheel rolling around a circle with twice its radius is a nephroid, illustrated in the centre, which looks like a pair of kidneys. The illuminated buttocks in your mug of tea when it's placed near a bright window are two half kidneys of a nephroid, since that is the shape produced by horizontal light reflecting off the inside of a circle, illustrated on the right.

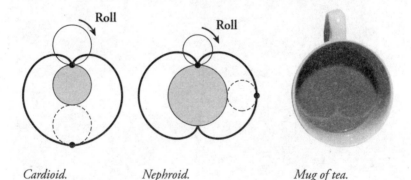

Cardioid.   *Nephroid.*   *Mug of tea.*

The first machine that drew curves for aesthetic as well as scientific reasons drew roulettes. The 'geometric pen', invented by the Italian Giambattista Suardi in the eighteenth century, consisted of a tripod with a rotating arm, on which a rotating cog held a pencil. 'There never was, perhaps, any instrument which delineates so many curves as the geometric pen,' gushed George Adams Jr., Instrument Maker to King George II. The designs were baroque and magical. In the nineteenth century P. H. Desvignes of Vienna designed a roulette-drawer he called the 'spirograph', which engraved the curves on a copper plate using a diamond stylus. The machine was used to create elaborate patterns for banknotes, to prevent forgeries. Drawing roulettes with a plastic set of cogs called the Spirograph, a toy launched in 1965, remains a rite of passage for nerdy kids.

One of my favourite pub maths puzzles involves rolling one coin around another. Put two identical coins next to each other flat on a table, as below, with the heads both upright. Roll the left coin around the right. What is the position of the head when the left coin reaches the right side?

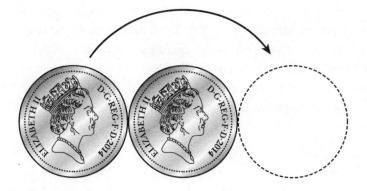

Rolling with money.

When I was first asked this question, I assumed that the rolling coin would be upside down, as it has only gone halfway around the fixed coin. I was wrong. The Queen completes a full revolution. It's a strange and counterintuitive experience to see her spin with regal speed, as if desperate to maintain her dignity in an upright position. The movement is a result of the property of all roulettes: they combine two independent directions of motion. In this case the coin is rotating around itself but also around the other coin. For every degree the left coin rolls around the right coin, it rotates two degrees around itself.

———

We generate roulettes by rolling a wheel. We can also generate curves by *spinning* a wheel with a fixed centre. These curves are mathematically simpler than roulettes, since we are only concerned with one direction of motion, the rotation about the centre.

Consider a point on the rim of a wheel spinning anticlockwise, illustrated below (i). If we plot its height relative to the angle of rotation when projected along a horizontal axis, we get a curve called the 'sinusoid' or 'sine wave'. I have marked the point on the circle at 0, 45, 90, 225 and 270 degrees. The sinusoid peaks at 90 degrees, returns to the horizontal axis at 180 degrees, then heads below the axis and is back where it started after a full turn. If the wheel keeps spinning, the curve will repeat once every rotation, painting symmetrical undulations ad infinitum.

(You may be wondering why this wiggly wave is named after the sine, the ratio between two sides of a right-angled triangle, since waves are nothing like triangles. It all makes sense, however, when we remember that the concept of sine comes from the circle in the first place: it is the half-chord. In the illustration (ii) below, we see this more clearly by creating a right-angled triangle in the wheel. If we let the hypotenuse be 1, then $\sin \alpha = \frac{\text{opposite}}{\text{hypotenuse}} = \frac{\text{height of point}}{1} = $ height of point.)

(i)

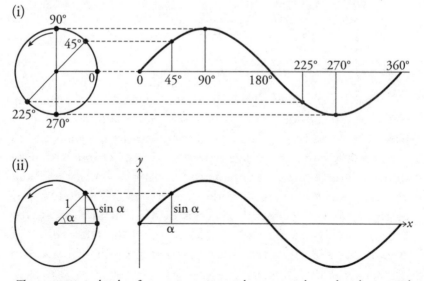

(ii)

The variation in height of a rotating point, with respect to the angle it has rotated, generates a sine wave.

The first person to draw a sinusoid was Gilles de Roberval in the seventeenth century. He called it the cycloid's 'companion curve'. The companion would later become pre-eminent in the hearts (and ears) of scientists and mathematicians.

The sinusoid is what's called a 'periodic wave', an entity in which a curve repeats itself again and again along the horizontal axis. The sinusoid is the simplest type of periodic wave because the circle, which generates it, is the simplest geometrical shape. Yet even though it is such a basic concept, the wave models many physical phenomena. The world is a carnival of sinusoids. The vertical position over time of a weight bouncing up and down on a spring is a sinusoid, as shown below left. The weight moves fastest in the middle of the oscillation, and slows down when it reaches the top and bottom, giving the recognizable shape (thinner here, since the horizontal scale is reduced). The horizontal position over time of a pendulum swinging side to side is also, for small swings, a sinusoid. Imagine a pendulum bob filled with fine sand that falls through a hole in the bottom, as shown below right. A bob swinging north–south leaving a deposit on a conveyor belt moving east–west will draw out a sine wave. Objects like the spring and the pendulum, which oscillate in sinusoids over time, are said to exhibit 'simple harmonic motion'.

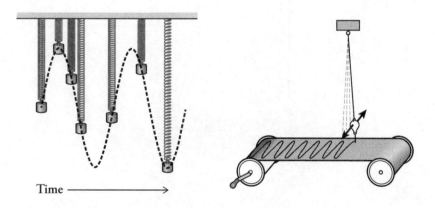

Time ⟶

A weight on a spring bounces, and a pendulum swings, with 'simple harmonic motion'.

121

We saw earlier that roulettes draw pretty patterns. So do sinusoids. In the 1840s, the Scottish mathematician Hugh Blackburn was experimenting with a sand-filled pendulum. He decided to hang the bob from two cords hanging in a Y, as illustrated below, joined by a ring at *r*. He held the ring still and swung the bob from left to right. Then he pulled the ring forwards and let it go, making it also swing forward and back. The bob was therefore being guided by two perpendicular swings, and the result was spectacular. The two competing sinusoidal motions pushed and pulled each other in a mathematical *pas de deux* that produced a sand pattern underneath of stunning intricacy. It was not long before enterprising instrument makers were manufacturing machines called harmonographs, in which two pendulums oscillated a pen simultaneously in two different directions. The user of the harmonograph would adjust the length of the pendulums, set the amplitude of their swings, and

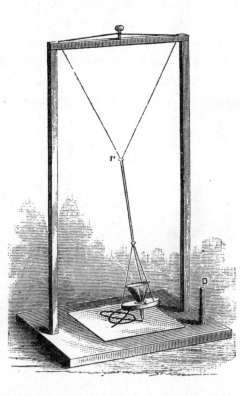

Blackburn's Y-shaped pendulum, from an 1879 popular science book.

then release them when the pen was positioned on a piece of paper. As the pen looped and swirled it traced beautiful geometrical shapes that were mechanical, yet somehow felt alive.

The Victorian harmonograph looked like a cross between a desk cabinet and a grandfather clock. Watching the pen create the images was hypnotic, a performance as much as a process. The loss of energy through friction, or 'damping', created curves that spiralled inwards as they headed towards the stationary point of equilibrium. Some of the larger machines were able to keep swinging for an hour or more before the pendulums came to a standstill.

The harmonograph was so popular that many other machines were manufactured based on the same principle: the sympalmograph, the pendulograph, the duplex pendulum, the quadruple harmonic-motion pendulum and, in the early twentieth century, the Creighton Compound Harmonic Motion Machine and the photo-ratiograph, which rendered the curves on photographic paper with a moving point of light. In the 1950s, the American artist John Whitney built a harmonograph out of Second World War military junk. He bought an M5 anti-aircraft director – a large metal box of cranks and levers that amounted to an early analogue computer used to calculate which way to fire at planes – and customised it so that the rotating parts could move a stylus with simple harmonic motion in two directions. Whitney could adjust the speed and size of the sinusoids electronically, giving him much more control and eliminating the effects of damping. The patterns he produced were dazzling and became some of the most iconic images in the history of mathematical art. They were famously used in the title sequence and posters for Alfred Hitchcock's 1958 movie *Vertigo*; the

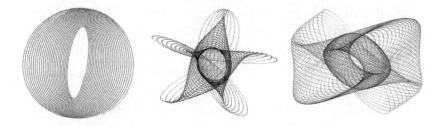

Good vibrations: figures produced by harmonographs.

swirling, dizzying concentric loops were a perfect visual metaphor for the film's tormented inner landscape. Not only were Whitney's patterns the first computer-generated images to feature in a Hollywood film, but his electronic harmonograph was also the first ever computer-animation machine.

Around the time that harmonographs became all the rage in Victorian drawing rooms, a physicist in Paris realized that he could generate identical figures using two tuning forks and a beam of light. The demonstrations by Jules Antoine Lissajous were some of the most beautiful experiments of the nineteenth century. When a tuning fork makes a sound, the metal prongs oscillate with simple harmonic motion. Lissajous attached a small mirror to one tuning fork and shone a beam at it so that it was reflected to a point on a screen. When the fork was made to vibrate, the point elongated into a horizontal line. The point was oscillating very fast back and forth, but the spectators saw it as a line because the image of the dot at each point persists in our vision for a split second longer than it is actually there. Lissajous then introduced a second tuning fork, which also had a mirror attached. The second fork was positioned at right angles to the first, so the light beam was reflected from the mirror on the first fork, oscillating in one direction, to the mirror on the second fork, oscillating in a perpendicular direction, and then onto the screen. The forks, in other words, were acting like the pendulums in a harmonograph, moving the beam in two competing sinusoidal oscillations. Yet instead of oscillating once every second or so, they were oscillating hundreds of times a second. Audiences saw striking images on the screen, which are now called Lissajous figures.

Different combinations of tuning forks produce different curves. When two identical tuning forks are played at the same volume, the sinusoids are equal and the curve looks like one from the first row of pictures shown opposite: an ellipse, a line or a circle. The particular curve depends on where each oscillation starts relative to the other. Lissajous adjusted this by changing the distance between the forks. When one fork oscillates at twice the frequency of the other, the curve is from the second line of images: a parabola or a figure of

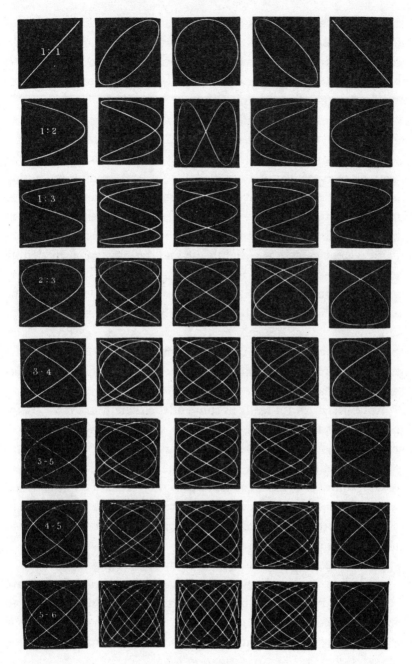

Lissajous figures, taken from a book published in 1875: for each row, the ratio between the frequencies of the sinusoids is marked in the left column.

eight. The remaining rows in the illustration show Lissajous figures for other whole-number ratios between the frequencies of the sinusoids. When the ratio of the frequencies cannot be described by two whole numbers, the beam will never return to the starting point and the image will be a blur.

The frequency of the oscillation of a tuning fork determines its note. A fork that oscillates at about 262 waves per second, for example, is middle C. Lissajous' experiments thus provided the music industry with an improved method for calibrating tuning forks. Rather than using your ear to hear whether two forks are in tune, you could use your eyes. Craftsmen set up beams of light in their workshops. If the pitches of the two forks were slightly off, their frequencies were also slightly off, and the curve made by doubly reflecting the beam was a mess. The technicians would use one fork as the standard, and then file down the other until the figure on the wall was an ellipse, indicating that the note was the same.

Lissajous figures are the result of combining two perpendicular sinusoidal oscillations. Can we combine sinusoids that are oscillating on the same axis?

Absolutely! And it leads to some of the richest and most useful theorems in mathematics. To help us along, let me define three concepts that are vital when talking about waves: frequency, amplitude and phase. Frequency is the number of times a wave oscillates in a set interval, amplitude is the vertical height between peak and trough, and phase is a measure of the horizontal position.

Armed with these concepts, we can outline an arithmetic of sinusoids, also illustrated opposite:

(i) This curve is the basic sine wave, with the equation $y = \sin x$.

(ii) When we double the frequency, which means that the wave repeats itself twice in the period that it takes the original curve to complete once, the equation becomes $y = \sin 2x$.

(iii) When we double the amplitude, which means that the wave oscillates twice as high, the equation is $y = 2 \sin x$.

(iv) When we alter the phase by shifting the wave a quarter of a wave to the left, we get the basic cosine wave, $y = \cos x$.

The waves produced by changing the frequency, amplitude and

phase of a sinusoid are all sinusoids. A helpful way to think about frequency, amplitude and phase is to remember that sinusoids are generated by a rotating point: the speed of the rotation determines the frequency, the radius of the rotating circle determines the amplitude and the point's starting position determines the phase.

(v) Here I have added the basic sine wave to the basic cosine wave. When we add two waves together, we just add the vertical values for each point on the horizontal axis. Magically, the result of adding the sine and cosine waves together is also a sinusoid, although one that has a different phase and an amplitude of $\sqrt{2}$. In fact, whenever two sinusoids with the same frequency are added together the result is always a sinusoid, whatever their amplitudes and phases.

In other words, a sinusoid added to any number of sinusoids that have the same frequency but different amplitudes and phases remains sinusoidal, like a science fiction monster that always remorphs into its original form. We'll return to the arithmetic of rotating points shortly, but first let's take a brief detour to another type of revolution, the French one.

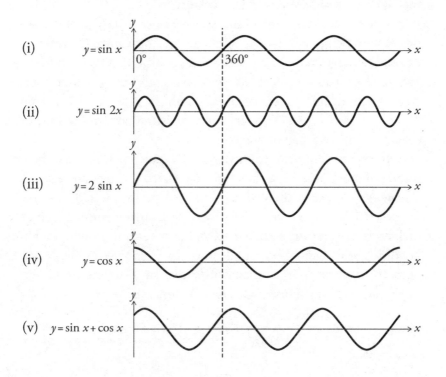

127

In 1798 Joseph Fourier, a 30-year-old professor at the École Polytechnique in Paris, received an urgent message from the Minister of the Interior informing him that his country required his services, and that he should 'be ready to depart at the first order'. Two months later, Fourier set sail from Toulon as part of a 25,000-strong military fleet under the command of General Napoleon Bonaparte, whose unannounced objective was the invasion of Egypt.

Fourier was one of 167 eminent scholars, the *savants*, assembled for the Egyptian expedition. Their presence reflected the French Revolution's ideology of scientific progress, and Napoleon, a keen amateur mathematician, liked to surround himself with colleagues who shared his interests. It is said that when the French troops reached the Great Pyramid at Giza, Napoleon sat in the shade underneath, scribbled a few notes in his jotter and announced that there was enough stone in the pyramid to build a wall three metres high and a third of a metre thick that would almost perfectly encircle France. Gaspard Monge, his chief mathematician, confirmed that the General's estimate was indeed correct.*

Fourier assumed many administrative roles in Egypt, the most prominent being permanent secretary of the Cairo Institute, a cultural heritage centre modelled on the Institut de France in Paris. When the Institute decided to collate all its scientific and archaeological discoveries, eventually published as the 37-volume *Description de L'Égypte*, Fourier wrote the introduction. He was, in effect, the father of Egyptology.

On Fourier's return from Egypt, Napoleon appointed him prefect of the Alpine department of Isère, based in Grenoble. Always a man of fragile health, with extreme sensitivity to cold, Fourier never left home without an overcoat, even in the summer, often making sure a servant carried a second coat for him in reserve. He kept his rooms baking hot at all times. In Grenoble, his academic research was also preoccupied with heat. In 1807 he published a ground-breaking paper, *On the Propagation of Heat in Solid*

---

* The Great Pyramid has sides of length 229m and a height of 146m. France is roughly a rectangle 770km north–south by 700km east–west. With these figures Napoleon's estimate is only 3 per cent off.

Bodies. In it he revealed a remarkable finding about sinusoids.

Fourier's famous theorem states that every periodic wave can be built up by adding sinusoids together. The result is surprising. Fourier's contemporaries met it with disbelief. Many waves look nothing at all like sinusoids, such as the square wave, illustrated below, which looks like the crenellations of a castle parapet. The square wave is made up of straight lines, whereas the sinusoid is continuously curved. Yet Fourier was right: we can build a square wave with only sinusoids.

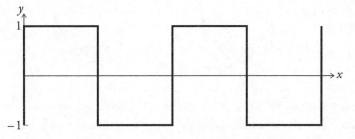

Here's how. In the illustration below there are three sine waves: the basic wave, a smaller sine wave with three times the frequency and a third of the amplitude, and an even smaller sine wave with five times the frequency and a fifth of the amplitude. We can write these three waves as $\sin x$, $\frac{\sin 3x}{3}$, and $\frac{\sin 5x}{5}$.

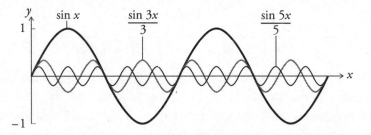

In the illustration overleaf I have started to add these waves together. We see the basic wave, $\sin x$. The sum $\sin x + \frac{\sin 3x}{3}$ is a wave that looks like a row of molar teeth. The sum $\sin x + \frac{\sin 3x}{3} + \frac{\sin 5x}{5}$ is a wave that looks like the filaments of a light bulb. If we carry on adding terms of the series:

$$\sin x + \frac{\sin 3x}{3} + \frac{\sin 5x}{5} + \frac{\sin 7x}{7} + \dots$$

we will get closer and closer to the square wave. At the limit, adding an infinity of terms, we will have the square wave. It is stunning that such a rigid shape can be constructed using only undulating wiggles. Any periodic wave consisting of jagged lines, smooth curves, or even a combination, can be built up with sinusoids.

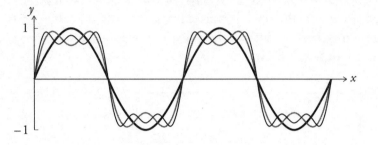

The sum of sinusoids that make up a wave is called its 'Fourier series'. It's a remarkably useful concept since it allows us to understand a continuous wave in terms of discrete signals. The terms in the series for the square wave, for example, can now be represented by the bar chart below.

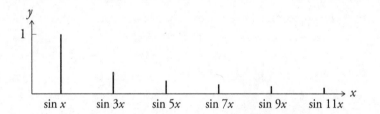

The horizontal axis represents the frequencies of the constituent sinusoids, and the vertical axis their amplitudes. Each bar stands for a sinusoid, and the leftmost bar is the sinusoid that has the 'fundamental' frequency. This type of graph is known as the 'frequency spectrum', or 'Fourier transform', of the wave.

Fourier's theorem was one of the most significant mathematical results of the nineteenth century because phenomena in many fields – from optics to quantum mechanics, and from seismology to electrical engineering – can be modelled by periodic waves. Often, the best way to investigate these waves is to break them down into simple sinusoids. The science of acoustics, for example, is essentially an application of Fourier's discoveries.

Sound is the vibration of air molecules. The molecules oscillate in the direction of travel of the sound, as illustrated overleaf with the clarinet, forming alternate areas of compression and rarefaction. The variation in air pressure at any point over time is a periodic wave.

As you can see, the clarinet wave is jagged and complicated. Fourier's theorem tells us, however, that we can break it down into a sum of sinusoids, whose frequencies are all multiples of the 'fundamental' frequency of the first term. In other words, the wave can be represented as a spectrum of frequencies with different amplitudes. The picture overleaf shows the frequency spectrum of the clarinet wave as a bar chart.

Remember, the jagged wave and the bar chart represent exactly the same sound, but in each image the information is encoded differently. For the wave, the horizontal axis is time, whereas on the bar chart the horizontal axis is frequency. Sound engineers say that the wave is in the 'time domain', and the transform is in the 'frequency domain'.

The frequency domain also provides us with all the information we need to recreate the sound of a clarinet using only tuning forks. Each bar in the bar chart represents a sinusoid oscillating at a fixed frequency. Now recall Lissajous' experiments with tuning forks from earlier in this chapter. The sound wave made by a tuning fork is a sinusoid. So, in order to recreate the sound of a clarinet, all that is required is to play a selection of tuning forks at the correct frequencies and amplitudes described by the bar chart. Likewise, the frequency spectrum of a violin would provide us with instructions on how to use tuning forks to produce the sound of a violin. The difference in timbre between middle C when played on the clarinet and the violin is the result of the same set of tuning forks oscillating with different relative amplitudes. A consequence of Fourier's theorem is that it is theoretically possible, therefore, to play the complete works of Beethoven with tuning forks, in such a way that is audibly indistinguishable from a symphony orchestra.

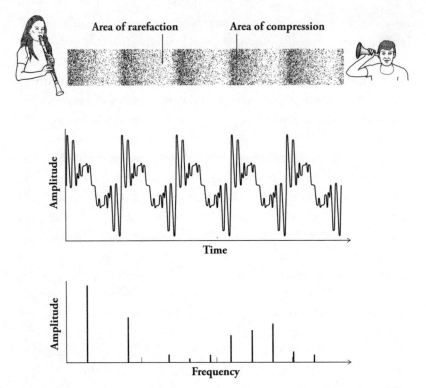

The sound wave and frequency spectrum of a clarinet.

When a fire engine passes Dolby Laboratories in San Francisco, everybody clasps their ears – especially the 'golden ears', those members of staff with exceptional hearing – hoping to protect their auditory faculties from deleterious noise. Dolby built its reputation on noise reduction systems for the music and film industries, and it now creates sound quality software for consumer electronic devices, using technology based entirely on sinusoids.

The benefit of being able to switch a sound wave from the time domain to the frequency domain is that some jobs that are really difficult in one domain become much simpler in the other. All sound played out of digital devices – such as your TV, phone and computer – is stored as data in the frequency domain, rather than the time domain. 'The waveform is like a noodle,' Brett Crockett, senior director of research sound technology, told me. 'You can't grab it.' Frequencies are much easier to store because they are a set of discrete values. It also helps that our ears cannot hear all frequencies. '[Ears] don't need the whole picture,' Crockett added. Dolby's software turns sound waves into sinusoids, and then strips out nonessential sinusoids so that the best possible sound can be recorded and stored with the least possible information. When the information is played back as sound, the spectrum of remaining frequencies is reconverted into a wave in the time domain.

It sounds easy, but in practice the task of filleting sinusoids from the frequency spectrum is exceedingly complex. It relies, firstly, on what's called a Fast Fourier Transform, a computer algorithm that converts the wave into its frequencies in real time. Secondly, different instruments, musical styles and voices require different solutions. One of the hardest sounds to get right is the harmonica, because its frequency spectrum looks like a picket fence – the amplitudes of the different frequencies are at the same height, forcing you to delete frequencies you can hear. For all Dolby's state of the art knowhow, the piece of music its software struggles most to recreate faithfully is *Moon River*, Henry Mancini's hauntingly beautiful 1961 song. Brett Crockett's golden ears judge new Dolby technology based on how faithfully it plays a harmonica riff recorded more than half a century ago.

Joseph Fourier was the first person to transform a periodic wave into a spectrum of frequencies. Much later, biologists figured out how the ear works. The part that hears is the cochlea, a tube full of liquid and lined with hair cells, coiled in the inner ear. The hairs vibrate according to the frequency of the incoming sound wave, with the lowest frequencies vibrating hairs at one end of the cochlea, and the highest frequencies vibrating hairs at the other. If we uncoiled the cochlea into a straight line it would look like the horizontal axis of a Fourier transform. Nature has been isolating the frequencies of sound waves for as long as creatures have had ears to hear.

This chapter has been about the maths of going round in circles. Let's break the cycle. What happens to things that get bigger and bigger and bigger?

All About *e*

In Boulder, Colorado, I visited the man who has given what is arguably the most performed lecture in the history of science. Albert Bartlett, emeritus physics professor at the University of Colorado, premiered his talk *Arithmetic, Population and Energy* in 1969. By the time I met him, he had given it 1712 times, and despite being almost ninety was continuing at a rate of about 20 times a year. Bartlett has a tall, robust physique and an imposing dome of a head, and he was wearing a Wild West bolo tie with stars and a planet emblazoned on its clasp. In his famous lecture, he proclaims with ominous portent that the greatest shortcoming of the human race is its inability to understand exponential growth. This simple but powerful message has in recent years propelled him to online stardom: a recording of his talk on YouTube, entitled *The Most IMPORTANT Video You'll Ever See*, has registered more than five million views.

Exponential (or proportional) growth happens when you repeatedly increase a quantity by the same proportion; by doubling, for example:

1, 2, 4, 8, 16, 32, 64 …

Or by tripling:

1, 3, 9, 27, 81, 243, 729 …

Or even increasing by just one per cent:

1, 1.01, 1.0201, 1.0303, 1.0406, 1.05101, 1.06152 …

We can rephrase these numbers using the following notation:

$2^0, 2^1, 2^2, 2^3, 2^4, 2^5, 2^6 \ldots$

$3^0, 3^1, 3^2, 3^3, 3^4, 3^5, 3^6 \ldots$

$1.01^0, 1.01^1, 1.01^2, 1.01^3, 1.01^4, 1.01^5, 1.01^6 \ldots$

The small number at the top right of the normal-sized number is called the 'exponent', or power, and it denotes the number of times you must multiply the normal-sized number by itself. Sequences that increase proportionately display 'exponential' growth – since for each new term the exponent increases by one.

When a number grows exponentially, the bigger it gets the faster it grows, and after only a handful of repetitions the number can reach a mind-boggling size. Consider what happens to a sheet of paper as you fold it. Each fold doubles the thickness. Since paper is about 0.1mm thick, the thickness in millimetres after each fold is

0.1, 0.2, 0.4, 0.8, 1.6, 3.2, 6.4 …

which is the doubling sequence we saw above, but with the decimal point one place along. Because the paper is getting thicker, each successive fold requires more strength, and by the seventh it is physically impossible to fold any further. The paper's thickness at this point is 128 times that of the single sheet, making it as thick as a 256-page book.

But let's carry on, to see – theoretically at least – how thick the piece of paper will get. After six more folds the paper is almost a metre tall. Six folds after that it is the height of the Arc de Triomphe, and six folds after that it towers 3km into the sky. Ordinary though doubling is, when we apply the procedure repeatedly, it does not take long for the results to be extraordinary. Our paper passes the Moon after 42 folds, and the total number required for it to reach the edge of the observable universe is just 92.

Albert Bartlett is less interested in other planets than the one we live on, and in his lecture he introduced a brilliantly compelling analogy for exponential growth. Imagine a bottle containing bacteria whose numbers double every minute. At 11am the bottle has one bacterium in it and an hour later, at noon, the bottle is full. Working backwards, the bottle must be half full at 11.59am, a quarter full at 11.58am, and so on. 'If you were an average bacterium in that bottle,' Bartlett asks, 'at what time would you realize you are running out of space?' At 11.55am the bottle looks pretty empty – it is only $\frac{1}{32}$ or about 3 per cent full, leaving 97 per cent free for expansion. Would the bacteria realize that they were only five minutes from full capacity? Bartlett's bottle is a cautionary tale about the Earth. If a population is growing exponentially, it will run out of space much sooner than it thinks.

Take the history of Boulder. Between 1950, the year Bartlett moved there, and 1970, the city's population rose by an average of six per cent a year. That's equivalent to multiplying the initial population by 1.06 to get the population at the end of year 1, multiplying the initial population by $(1.06)^2$ to get the population at the end of year 2, multiplying the initial population by $(1.06)^3$ to get the population at the end of year 3, and so on, which is clearly an exponential progression.

Six per cent a year doesn't sound that much on its own, but after two decades it represents more than a tripling of the population, from 20,000 to 67,000. 'That is just a staggering amount,' said Bartlett, 'and we have been fighting ever since to try and slow it down' (it is now almost 100,000). Bartlett's passion for explaining exponential growth came from a determination to preserve the quality of life in his mountainside hometown.

It is important to remember that whenever the percentage increase per unit of time is constant, growth is exponential, which means that even though the quantity under discussion may start growing slowly, growth will soon speed up and before long the quantity will be counter-intuitively huge. Since almost all economic, financial and political measures of growth – of sales, profits, stock prices, GDP and population, for example, as well as inflation and interest rates – are calculated as a percentage change per unit of time, exponential growth is crucial to understanding the world.

Which was also true half a millennium ago, when concern about exponential growth led to the widespread use of an arithmetical rule of thumb: the Rule of 72, first mentioned in Luca Pacioli's *Summa de Arithmetica*, the mathematical bible of the Renaissance. If a quantity is growing exponentially then there is a fixed period in which it doubles, known as the 'doubling time'. The Rule of 72 states that a quantity growing at X per cent every time period will double in size in roughly $\frac{72}{X}$ periods. (I explain how this is worked out in Appendix Five on p. 298). So, if a population is growing at 1 per cent a year it will take about $\frac{72}{1}$, or 72 years, for the population to double. If the town is growing at two per cent a year it will take $\frac{72}{2}$, or about 36 years, and if it is growing at six per cent, as it was in Boulder, it will take $\frac{72}{6}$, or 12 years.

Doubling time is a useful concept because it allows us to see easily into the future and the past. If Boulder's population doubles in 12 years, it will quadruple in 24 years, and after 36 years it will be eight times as big. (Providing the growth rate stays the same, of course.) Likewise, if we think about previous growth, a quantity growing at six per cent a year would have had half its current value 12 years ago, a quarter of its current value 24 years ago, and 36 years ago it would have been an eighth of the size.

By converting a percentage to a doubling time you get a better feel for how fast the quantity is accelerating, which makes the Rule of 72 indispensable when thinking about exponential growth. I remember my father explaining it to me when I was young, and he was taught it by his father, who as a clothes trader in the East End of London before the era of calculators would have relied on it in his working life. If you borrow money at 10 per cent annual interest, the rule tells you with little mental effort that the debt will double in about seven years, and quadruple in fourteen.

Albert Bartlett's interest in exponentials soon grew beyond issues of overcrowding, pollution and traffic congestion in Boulder, since the same arguments he made to the city council applied equally to the world as a whole. The Earth cannot sustain a global population that is growing proportionately every year, or not for much longer anyway. How many minutes left are there, he asked, before the bacteria fill the bottle? Bartlett's views have made him a modern-day Thomas Malthus, the English cleric who two hundred years ago argued that population growth leads to famine and disease, since the exponential growth of people cannot be matched by a corresponding growth in food production. 'Malthus was right!' Bartlett said. 'He didn't anticipate petroleum and mechanization, but his ideas were right. He understood exponential growth versus linear growth. The population has a capability of growing more rapidly than we can grow any of the resources we need to survive.' And he added: 'No matter what assumptions you make, the population crashes in the middle of this century, about 40 years from now.'

Bartlett is a gripping lecturer. He cleverly converts the vertigo you feel when considering exponential growth into the fear of an imminent, apocalyptic future. His talk is also refreshing in the way he brings the tools of physics – distilling the essence of the problem, isolating the universal law – to a discussion usually dominated by economists and social scientists. He reserves much of his ire for economists, whom he blames for a collective denial. 'They have built up a society in which you have to have population growth to have job growth. But growth never pays for itself, and leads to disaster ultimately.' The only viable option, he said, is for society to break its addiction to exponentials.

Opponents of Bartlett's view argue that science will come up with solutions to increase production of food and energy, as it has managed to until now, and that birth rates are reducing globally anyway. They miss the point, he said. 'A common response by economists has been that I don't understand the problem, that it is more complex than this simple-minded thing. I reply that if you don't understand the simple aspects, you can't hope to understand the complicated aspects!' And then he chuckled: 'That doesn't get me anywhere. You can't sustain population growth, or growth in the consumption of resources, period. End of argument. It isn't debatable, unless you want to debate arithmetic.'

Bartlett calls our inability to understand exponential growth the greatest shortcoming of the human race. But why do we find it so hard to comprehend? In 1980 the psychologist Gideon Keren of the Institute for Perception in the Netherlands conducted a study that attempted to see if there were any cultural differences in misperceptions of exponential growth. He asked a group of

Canadians to predict the price of a piece of steak that was rising 13 per cent year on year. The subjects were told the price in 1977, 1978, 1979 and 1980, when it was $3, and they had to estimate the value it would reach 13 years later in 1993. The average guess was $7.7, about half the value of the correct answer, $14.7, which was a significant shortfall. Keren then asked the same question of a group of Israelis using their local currency, Israeli pounds, and with the 1980 price of a steak being I£25. The average guess for the price in 1993 was I£106.4, which while again an underestimate – the correct answer was I£122.4 – was much closer to the target. Keren argued that the Israelis performed better because their country was going through a period of about 100 per cent annual inflation at the time of the survey, compared to about 10 per cent in Canada. He concluded that with more experience of higher exponential growth, Israelis had become more sensitive to it, even if their estimates still undershot.

In 1973 Daniel Kahneman and Amos Tversky demonstrated that people make much lower estimates when evaluating $1 \times 2 \times 3 \times 4 \times 5 \times 6 \times 7 \times 8$ than they do for $8 \times 7 \times 6 \times 5 \times 4 \times 3 \times 2 \times 1$, which are identical in value, showing that we are unduly influenced by the order in which we read numbers. (The median response for the rising sequence was 512, and 2250 for the descending one. In fact, both estimates were drastically short of the correct answer, 40,320.) Kahneman and Tversky's research provides an insight into why we will always underestimate exponential growth: in any sequence we are subconsciously pulled down, or 'anchored', by the earliest terms, and the effect is most extreme when the sequence is ascending.

Exponential growth can be step-by-step, or continuous. In Bartlett's bacteria-in-the-bottle analogy, one bacterium becomes two, two bacteria become four, four become eight, and so on. The population increases by whole numbers at fixed intervals. The curves in the illustration opposite, however, are growing exponentially *and* continuously. At every point the curve is rising at a rate proportional to its height.

When an equation is of the form $y = a^x$, where a is a positive number, the curve displays continuous exponential growth. The

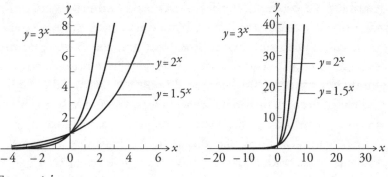

Exponential curves.

curves above are described by the equations $y = 3^x$, $y = 2^x$, and $y = 1.5^x$, which are the continuous curves of the tripling sequence, the doubling sequence and the sequence where each new term grows by 50 per cent. In the case of $y = 2^x$, for example, when the values of x are 1, 2, 3, 4, 5 …, the values of y are 2, 4, 8, 16, 32 …

On the smaller-scale graph, left, the curves look like ribbons pinned to the vertical axis at 1. On the larger scale graph, right, you can see how they all suffer similar fates – becoming close to vertical only a few units along the x axis. It doesn't look like these curves will eventually cover the plane horizontally, although inevitably they will. If I wanted the graph on the right to show $y = 3^x$, where $x = 30$, the page would need to stretch for more than a hundred million kilometres in the vertical direction.

When a curve rises exponentially, the higher it gets the steeper it becomes. The further we travel up the curve, the faster it grows. Before we continue, however, we need to familiarize ourselves with a new concept: 'gradient', the mathematical measure of steepness. The gradient of a slope is $\frac{\text{change in height}}{\text{change in distance}}$ and should already be familiar to anyone who has driven or cycled up a mountain road. If a road rises 100m while travelling 400m horizontally, as illustrated below, then the gradient is $\frac{100}{400}$, or $\frac{1}{4}$, which a road sign will usually

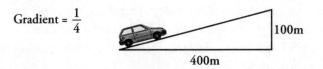

Gradient $= \dfrac{1}{4}$

100m

400m

describe as 25 per cent. The definition makes intuitive sense, since it means that steeper roads have higher gradients. But we need to be careful. A road that has 100 per cent gradient is one that rises the same amount as it travels across, meaning that it rises at an angle of only 45 degrees. It is theoretically possible for a road to have a gradient of more than 100 per cent; in fact, it could have a gradient of infinity, which would be a vertical slope.

The road illustrated on the previous page has a constant gradient. Most roads, however, have a variable gradient. They get steeper, flatten out, get steeper again. To find the gradient at a point on a road, or curve, with changing steepness, we must draw the tangent at that point and find *its* gradient. The tangent is the line that touches the curve at that point but does not cross it (the word 'tangent' comes from the Latin *tangere*, to touch). In the illustration below of a curve with varying gradient, I have marked a point P and drawn its tangent. To find the gradient of the tangent, we draw a right-angled triangle, which tells us the height change, a, for the horizontal change, b, and then calculate $\frac{a}{b}$. The size of the triangle doesn't matter, because the ratio of height to width will stay the same. The gradient at point P is the gradient of the tangent at P, which is $\frac{a}{b}$.

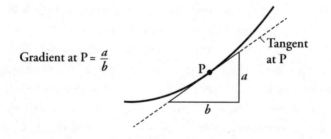

Gradient at P = $\frac{a}{b}$

Let's return to my description of exponential curves: the further we travel up them, the steeper they get. In other words, the higher up the curve you go, the higher the gradient. In fact, we can make a bolder statement. For all exponential curves, *the gradient is always a fixed percentage of the height*. Which leaves us with an obvious question. What is the Goldilocks curve, where the gradient and the height are always equal?

The 'just right' curve turns out to be:

$$y = (2.7182818284...)^x$$

When the height is 1, the gradient is 1, when the height is 2, the gradient is 2, and when the height is 3, the gradient is 3, as illustrated in the triptych below. Correspondingly, when the height is pi, the gradient is pi, and when the height is a million, the gradient is a million. As you move along the curve, its two most fundamental properties – height and gradient – are equal and rise together, an ascent of joyous synchronicity, like lovers in a Chagall painting floating up into the sky.

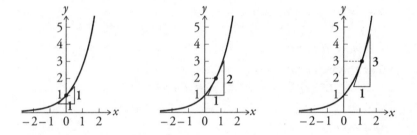

The curve of y = e^x: the height of a point on the curve is always equal to the gradient at that point.

The geometric beauty of the curve, however, contrasts with its ugly spawn: a splutter of decimal digits, beginning 2.718 and continuing for ever without repeating themselves. For convenience we represent this number with the letter *e*, and call it the 'exponential constant'. It is the second-most-famous mathematical constant after pi. Unlike pi, however, which has been the subject of interest for millennia, *e* is a Johnny-come-lately.

When asked what was the greatest invention of all time, Albert Einstein is said to have quipped: 'compound interest'. The exchange probably never took place, but it has entered urban mythology because it is just the type of playful answer we like to think he would have made. Interest is the fee you pay for borrowing money or what you receive for lending it. It's usually a percentage of the money borrowed or lent. Simple interest is money paid on the original amount, and it stays the same each instalment. So if a bank charges 20 per cent annual simple interest on a loan of £100, then after one year the debt is £120, after two years it is £140, after three years it is £160, and so on. With compound interest, however, each payment is a proportion of the *compounded*, or accumulated, total. So, if a bank charges 20 per cent annual compound interest, a £100 debt will become £120 after a year, £144 after two, and £172.80 after three, because:

Year one: debt + interest
$$= £100 + (£100 \times \tfrac{20}{100}) = £120$$

Year two: accumulated debt + interest
$$= £120 + (£120 \times \tfrac{20}{100}) = £144$$

Year three: accumulated debt + interest
$$= £144 + (£144 \times \tfrac{20}{100}) = £172.80$$

And so on.

Compound interest grows much faster than simple interest because it grows exponentially. Adding X per cent to a debt is equivalent to multiplying it by $(1 + \tfrac{X}{100})$, so the calculation above is equally:

Year one: $£100 \, (1 + \tfrac{20}{100})$

Year two: $£100 \, (1 + \tfrac{20}{100}) \times (1 + \tfrac{20}{100}) = £100 \, (1 + \tfrac{20}{100})^2$

Year three: $£100 \, (1 + \tfrac{20}{100})^2 \times (1 + \tfrac{20}{100}) = £100 \, (1 + \tfrac{20}{100})^3$

Which is a sequence growing exponentially.

Moneylenders have preferred compound over simple interest for as long as we know. Indeed, one of the earliest problems in mathematical literature, on a Mesopotamian clay tablet dating from 1700 BCE, asks how long it will take a sum to double if the interest is compounded at 20 per cent a year. One reason why banking is so lucrative is that compound interest increases the debt, or loan, exponentially, meaning that you can end up paying, or earning, exorbitant amounts in a short time. The Romans condemned compounding as the worst form of usury. In the Koran it is decreed sinful. Nevertheless, the modern global financial system relies on the practice. It's how our bank balances, credit card bills and mortgage payments are calculated. Compound interest has been the chief catalyst of economic growth since civilization began.

In the late seventeenth century the Swiss mathematician Jakob Bernoulli asked a pretty basic question about compound interest. In what way does the compounding interval matter to the value of a loan? (Jakob was the elder brother of Johann, who we met last chapter challenging the world's most brilliant mathematicians to find the path of quickest descent.) Is it better to be paid the full interest rate once a year, or half the annual rate compounded every half a year, or a twelfth the annual rate compounded every month, or even $\frac{1}{365}$ the annual rate compounded every day? Intuitively, it would seem that the more we compound the more we stand to make, which is indeed the case, since the money spends more time 'working' for us. I will take you through the calculation step by step, however, since it throws up an interesting arithmetical pattern.

To keep the numbers as simple as possible, let's say our deposit is £1 and the bank pays an interest rate of 100 per cent per annum. After one year, the value of the deposit will double to £2.

If we halve the interest rate and the compounding interval, we get a 50 per cent rate to be compounded twice.

After six months our deposit will grow to:

$$£(1 + \tfrac{50}{100}) = £1.50$$

And after a year it will be:

$$£(1 + \tfrac{50}{100})(1 + \tfrac{50}{100}) = £(1 + \tfrac{50}{100})^2 = £(1 + \tfrac{1}{2})^2 = £2.25$$

So, by compounding every six months, we make an extra 25p.

Likewise, if the interest rate is a 12th of 100 per cent and there are twelve monthly payments, the deposit will grow to:

$$£(1 + \tfrac{1}{12})^{12} = £2.6130$$

By compounding monthly, we make an extra 61p.

And if the interest rate is a 365th of 100 per cent, and there are 365 daily payments, the deposit will grow to:

$$£(1 + \tfrac{1}{365})^{365} = £2.7146$$

We make an extra 71p.

The pattern is clear. The more compounding intervals there are, the more our money earns. Yet how far can we push it? Jakob Bernoulli wanted to know if there were any limits to how high the sum could grow if the compounding intervals continued to divide into smaller and smaller time periods.

As we have seen, if we divide the annual percentage rate by n and compound it n times, the end-of-year balance in pounds is:

$$(1 + \tfrac{1}{n})^n$$

Rephrasing Bernoulli's question algebraically, he was asking what happens to this term as n approaches infinity. Does it also increase to infinity or does it approach a finite limit? I like to visualize this problem as a tug of war along the horizontal axis of a graph. As n gets bigger, $(1 + \tfrac{1}{n})$ gets smaller, pulling the term to the left. On the other hand, the exponent ^{n} is pulling the term to the right, since the more times you multiply what's in the brackets the larger the sum will be. At the start of the race the exponent is winning, since we have seen that when n is 1, 2, 12 and 365 the value of $(1 + \tfrac{1}{n})^n$ grows, from 2 to 2.25 to 2.6130 to 2.7146. You can probably see where we're going. When n heads to infinity, the tug-of-war reaches equilibrium. Bernoulli had unintentionally stumbled on the exponential constant, since in the limit, when n tends to infinity, $(1 + \tfrac{1}{n})^n$ tends to e.

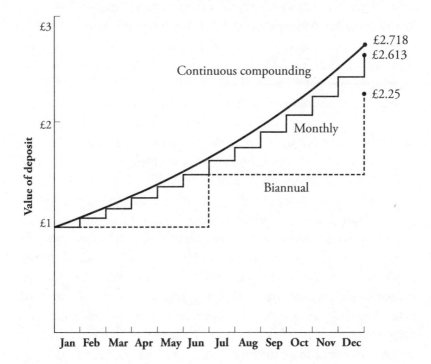

The value of £1 over a year when a 100 per cent annual interest rate is compounded biannually, monthly and continuously.

Let's look at this process visually. The illustration above contains three scenarios of what happens to a deposit of £1 over a year with a 100 per cent per annum interest rate, compounded proportionally at different intervals. The dashed line represents biannual compounding, and the thin line monthly compounding. With more steps, the lines rise higher. When the steps are infinitely small, the line is the curve $y = e^x$, the poster boy of exponential growth.

We say that the curve is 'continuously compounding', meaning that the value of our deposit is growing at every instant throughout the year. At the end of one year, the balance is £2.718…, or £*e*.

Bernoulli discovered *e* while studying compound interest. He would no doubt be delighted to discover that his foundling is the bedrock of modern banking (with more realistic interest rates, obviously). This is because British financial institutions are legally obliged to state the continuously compounded interest rate on every product they sell, irrespective of whether they choose to pay monthly, biannually, annually, or whatever.

Let's say that a bank offers a deposit account that pays 15 per cent per annum, compounded in one annual instalment, which means that after one year a £100 loan will grow to £115. If this 15 per cent is compounded continuously, a formula derived from the properties of e tells us that after one year the loan will grow to $£100 \times e^{15/100}$, which works out at £116.18, or an annual interest rate of 16.18 per cent. The bank is required by law to declare that this particular deposit account pays 16.18 per cent. While it seems odd for banks to declare a figure they don't use in practice, the rule was introduced so that customers could compare like with like. An account that pays monthly and one that pays annually are both judged by their continuously compounded rates. Since almost every financial product involves compound interest, and every calculation of continuous compounding throws up an e, the exponential constant is the pivotal number on which the entire financial system depends.

That's enough about money. Many other phenomena exhibit exponential growth, such as the spread of disease, the proliferation of microorganisms, the escalation of a nuclear chain reaction, the expansion of internet traffic and the feedback of an electric guitar. In all these cases, scientists model the growth using e.

I wrote previously that the equation $y = a^x$, where a is a positive number, describes an exponential curve. We can rephrase this equation so it has an e in it. The arithmetic of exponents means that the term a^x is equivalent to the term e^{kx} for a unique, positive number k. For example, the curve of the doubling sequence has the equation $y = 2^x$, but it can also be written $y = e^{0.693x}$. Likewise, the curve of the tripling sequence, $y = 3^x$, has the equivalent form $y = e^{1.099x}$. Mathematicians prefer to convert the equation $y = a^x$ into an equation $y = e^{kx}$ because e represents growth in its purest form. It simplifies the equation, facilitates calculation and is more

682   791

Girl or boy? When this photograph was paired with even digits, respondents thought it more likely that the baby was female. When it was paired with odd digits, they thought it more likely that the baby was male. (See pp. 5–6.)

Hold the front page! Gregory A. Gadawski and Darrell D. Dorrell, partners in an Oregon financial investigations firm, with local coverage of a fraudster they helped jail using Benford's law. (See pp. 31–3.)

At noon on midsummer's day, the bottom of this well near Aswan in Egypt is lit completely, because the sun is directly overhead. It's possible that the well was the one used by Eratosthenes in his measurement of the Earth. (See pp. 59–61.)

ABOVE: *Four men measuring the angles to faraway points, from a sixteenth-century German book on cartography. Heights and distances will then be calculated using trigonometry, enabling the drawing of accurate maps. (See pp. 73–4.)*

RIGHT: *Rob Woodall is the Bhagwan of 'trig baggers', the name given to ramblers who strive to visit as many triangulation pillars as possible (see p. 57). His conquests include this one, at Mount Ararat in Turkey.*

Iconic section: playing Elliptipool – pocket billiards with an ellipse-shaped table – was once the cool way to impress the opposite sex, as evidenced by this photograph, published in the interiors bible House & Garden *in 1964.*
(See pp. 81–2.)

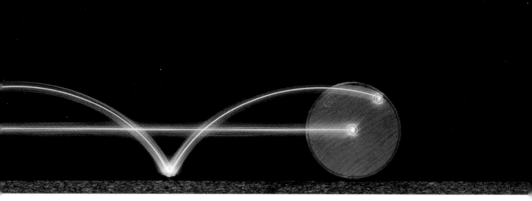

A light on the rim of a rolling wheel draws a cycloid (see pp. 111–12), above. This picture was taken by Berenice Abbott, a celebrated architectural photographer, who in 1958 was commissioned to illustrate a high school physics textbook. The photo below of a bouncing ball also came from this project. Each bounce is a parabola, and the height of each bounce decays exponentially (see pp. 91–2 and 149).

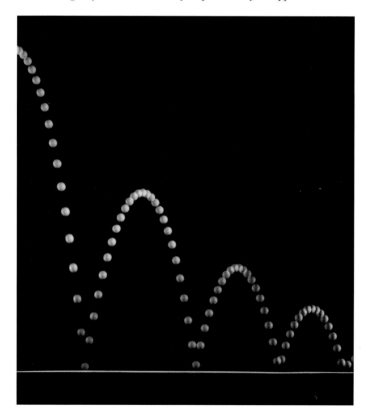

Designed by Théodore Olivier and manufactured in Paris in 1872, this string model, above left, shows how the curved surface of a hyperboloid is made out of straight lines. The hyperboloid-shaped Canton Tower in Guangzhou, below left, is made out of straight girders. (See pp. 101–4.)

OPPOSITE: *American artist John Whitney drew the swirling pattern for the 1958 film poster of* Vertigo *using a machine he made from Second World War anti-aircraft weaponry. It was the birth of computer art. (See pp. 123–4.)*

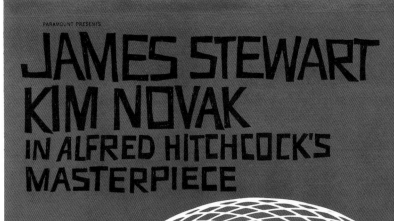

PARAMOUNT PRESENTS

JAMES STEWART
KIM NOVAK
IN ALFRED HITCHCOCK'S
MASTERPIECE

'VERTIGO'

CO STARRING
BARBARA BEL GEDDES WITH TOM HELMORE HENRY JONES DIRECTED BY ALFRED HITCHCOCK SCREENPLAY BY ALEC COPPEL & SAMUEL TAYLOR TECHNICOLOR® VISTAVISION®
BASED UPON THE NOVEL 'D'ENTRE LES MORTS' BY PIERRE BOILEAU AND THOMAS NARCEJAC MUSIC BY BERNARD HERRMANN

These instruments represent the two most popular families of mathematical drawing machines. Above is the Geometric Chuck, made in the early nineteenth century, which was an antecedent of the modern-day Spirograph (see pp. 118–19). It produces curves of superimposed circular motions, such as the image at top right. Left is a 1909 harmonograph, in which both pen and paper are attached to swinging pendulums. It produces Lissajous figures, such as the one at bottom right (see pp. 122–6).

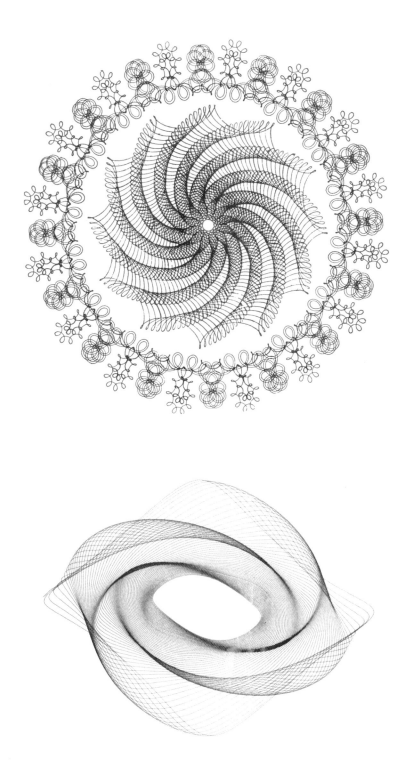

A piece of string hanging between two points naturally falls in the shape of the catenary curve, while a weighted string hangs in the shape of a 'transformed' catenary. When turned upside down these curves make robust arches. Between 1898 and 1908, the Catalan architect Antoni Gaudí made the hanging model of the Colònia Güell church above. The structure, made from string and sachets of lead shot, is dressed with material and ready to be photographed – turn the page around by 180 degrees to see the shape of the intended building. (See pp. 150–5.)

Catenaries in concrete and silk: plans for Kuwait airport (see p. 155), above, and a spider's web, below.

The Mandelbulb, above, is a three-dimensional representation of the Mandelbrot set, a fractal shape made by adding and multiplying complex numbers. A detail of the top of it is shown below. British fractal artist Daniel White produced the first images of the Mandelbulb in 2009. When he looked inside it, and zoomed in to a 26th of the overall size, he found the spine-like shape opposite. (See pp. 195–200.)

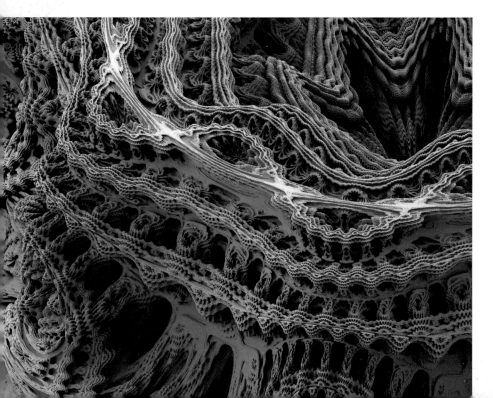

Star mathmo: Cédric Villani won the Fields Medal for his work on the behaviour of particles that fly through the air. His signature metal spider crouches on his left lapel. (See pp. 203–5 and 220–4.)

The calculus of cookies: Gottfried Leibniz is remembered for his feud with Isaac Newton and for the butter biscuit named in his honour. Coincidentally, the Newton is also a traditional baked snack. (See pp. 217–19.)

Vannevar Bush, pictured left, standing at the helm of his 'differential analyser', a machine built to solve differential equations, at the Massachusetts Institute of Technology in the early 1930s. (See p. 225.)

Members of Bourbaki, the secret French maths society, at their 1951 summer congress in the Alps. From left to right: Jacques Dixmier, Jean Dieudonné, Pierre Samuel, André Weil, Jean Delsarte and Laurent Schwartz. (See pp. 250–3.)

John von Neumann (left) and his best friend Stanislaw Ulam (right) devised the cellular automaton. This photograph was taken in 1949 near Los Alamos, where together with Richard Feynman (centre) they worked on the development of nuclear weapons. (See pp. 259–61.)

elegant. The exponential constant *e* is the essential element of the mathematics of growth.

Pi is the first constant we learn at school, and only those who specialize in maths will later learn *e*. Yet at university level, *e* pre-dominates. Even though it is purely coincidental that *e* is also the most common letter in the English language, the mathematical role of *e* actually has a parallel to its linguistic one. When an equation has an *e* in it, it indicates a bud of exponential growth, a flowering, a sign of life. Correspondingly, an *e* brings vitality to written language, trans-forming contiguous consonants into pronounceable words.

Exponential growth has an opposite: exponential decay, in which a quantity decreases repeatedly by the same proportion.

For example, the halving sequence:

$$1, \frac{1}{2}, \frac{1}{4}, \frac{1}{8}, \frac{1}{16}, \frac{1}{32} \cdots$$

exhibits exponential decay.

The equivalent concept of 'doubling time' for exponential decay is 'half-life', the fixed length of time it takes for a quantity to decrease by half. It's a common term in nuclear physics, for example. The number of radioactive particles in a radioactive material decays exponentially, and with enormous variation too: the half-life of hydrogen-7 is 0.00000000000000000000023 seconds, while for calcium-48 it is 40,000,000,000,000,000,000 years.

More mundanely, the difference in temperature between hot tea and the mug you have just poured it into decays exponentially, as does the atmospheric pressure as you walk up a hill.

The purest curve of exponential decay is $y = \frac{1}{e^x}$, which is also written $y = e^{-x}$, illustrated overleaf, for which the gradient is always the negative of the height. The decay curve is just the exponential curve $y = e^x$ reflected in the vertical axis. One interesting property of this curve is that the (shaded) region bounded by the curve, the vertical axis and the horizontal axis has finite area, equal to 1, even though the area is infinitely long, since the curve never reaches the horizontal axis.

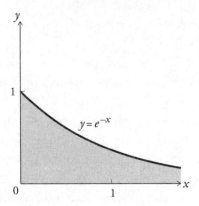

The curve $y = \frac{1}{e^x}$ of exponential decay.

In the May 1690 issue of the *Acta Eruditorum*, Jakob Bernoulli, the discoverer of *e*, revisited a question that had been puzzling mathematicians for a century. What is the correct geometry of the shape made by a piece of string when it is hanging between two points? This curve – called the 'catenary', from the Latin *catena, chain* – is produced when a material is suspended by its own weight, as shown on the opposite page. It's the sag of an electricity cable, the smile of a necklace, the U of a skipping rope and the droop of a velvet cord. The cross section of a billowing sail is also a catenary, rotated by 90 degrees, since wind acts horizontally as gravity does vertically. In a departure from many of the other mathematical challenges posed in the seventeenth century, however, Jakob did not know the answer to his question before he asked it. After a year's work it still eluded him. His younger brother Johann eventually found a solution, which you might assume would be a cause for great joy in the Bernoulli household. It wasn't. The Bernoullis were the most dysfunctional family in the history of mathematics.

Originally from Antwerp, the Bernoullis had fled the Spanish persecution of the Huguenots, and by the early seventeenth century were spice merchants settled in Basel, Switzerland. Jakob, born in 1654, was the first mathematician of what was to become a family dynasty unmatched in any field. Over three generations, eight Bernoullis would become distinguished mathematicians, each with significant discoveries to their names. Jakob, as well as studying compound interest, is perhaps best known for writing the first major

Maths bling: the catenary curve.

work on probability. He was also 'self-willed, obstinate, aggressive, vindictive, beset by feelings of inferiority and yet firmly convinced of his own abilities', according to one historian. This set him on a collision course with Johann, thirteen years his junior, who was similarly disposed. Johann relished solving the catenary problem, and later recounted the episode with glee: 'The efforts of my brother were without success; for my part, I was more fortunate, for I found the skill (I say it without boasting, why should I conceal the truth?) to solve it in full,' he wrote, adding: 'It is true that it cost me study that robbed me of rest for an entire night …' One *entire* night for a problem his brother failed to solve in a year? Ouch! Johann was as competitive with his sons as he was with his brother. After he and his middle son Daniel were awarded a shared prize from the Paris Academy, he was so jealous he barred Daniel from the family home.

The curve whose identity Jakob Bernoulli so ardently sought turned out to have a hidden ingredient, *e*, the number he had uncovered in a different context.

In modern notation, the equation for the catenary is:

$$y = \frac{e^{ax} + e^{-ax}}{2a}$$

where *a* is a constant that changes the scale of the curve. The bigger *a* is, the further apart the two ends of the hanging string are, as illustrated overleaf.

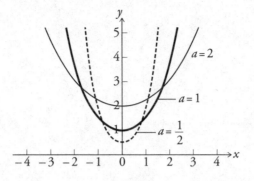

The catenary equation $\dfrac{e^{ax} + e^{-ax}}{2a}$ with different values for a.

If we let $a = 1$ in the catenary equation, then the curve is

$$y = \frac{e^x + e^{-x}}{2}$$

Look closely at the equation. The term e^x represents pure exponential growth, and the term e^{-x} pure exponential decay. The equation adds them together, and then divides by two, which is a familiar arithmetical operation; adding two values and halving the result is what we do when we want to find their average. In other words, the catenary is the average of the curves of exponential growth and decay, as illustrated on the opposite page. Every point on the U falls exactly halfway between the two exponential curves.

Whenever we see a circle, we see pi, the ratio of the circumference to the diameter. And whenever we see a hanging chain, a dangling spider's thread or the dip of an empty washing line, we see e.

In the seventeenth century, the English physicist Robert Hooke discovered an extraordinary mechanical property of the catenary: the curve, when upside down, is the most stable shape for a free-standing arch. When a chain hangs, it settles in a position where all its internal forces are pulling along the line of the curve. When the catenary is turned upside down, these forces of tension become forces of compression, making the catenary the unique arch where all compression forces are acting along the line of the curve. There are no bending forces in a catenary: it supports itself under its own

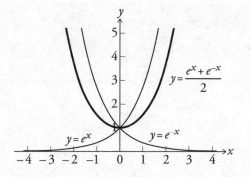

The catenary is the average of exponential growth and decay.

weight, needing no braces or buttresses. It will stand sturdily with a minimal amount of masonry. Bricks in a catenary will not even need mortar to stand stable, since they push against each other perfectly along the curve. Hooke was pleased with his discovery, declaring it an idea that 'no Architectonik Writer hath ever yet attempted, much less performed'. It was not long, however, before engineers were using catenaries. Before the computer age the quickest way to make one was to hang a chain, trace out the curve, build a model using a rigid material and stand it upside down.

The catenary is nature's legs, the most perfect way to stand on two feet. The arch, in fact, was a signature motif of Antoni Gaudí, the Catalan architect responsible for some of the most stunning buildings of the early twentieth century, notably the basilica of the Sagrada Família in Barcelona. Gaudí was drawn not just to the aesthetics of the catenary but also to what it represented mathematically. His use of catenaries made the structural mechanics of a building a principal feature of its design.

Arches in buildings, however, are rarely free-standing. Usually they form columns or vaults linked to walls, floors and the roof. Gaudí realized that the entire architecture of a building could be drafted using a model of hanging chains, and this is just what he did. For example, when he was commissioned to design a church at the Colònia Güell near Barcelona, he made an upside-down skeleton of the project. Instead of using metal chains, he used string weighed down by hundreds of sachets containing lead shot. The

A replica of a Gaudí chain model hanging in the Pedrera-Casa Milà museum in Barcelona. Turn the page upside down to see the shape of the intended building.

weight of each sachet on the string created a mesh of 'transformed' catenary curves. The arches of these transformed catenaries were the most stable curves to withstand a corresponding weight at the same position (such as the roof, or building materials). To see what the finished church would look like, Gaudí took a photograph and turned it upside down. Even though the Colònia Güell Church was never finished, it inspired him to use the same technique in his later work.

The most famous catenary is probably the Gateway Arch in St Louis, Missouri, which is 192m high, although it is slightly flatter than a perfect curve to take account of thinner masonry at the top. In 2011 London architects Foster and Partners decided on the catenary for a particularly challenging project: a mega-airport in Kuwait, one of the most inhospitable but inhabited places on Earth. Nikolai Malsch, the lead architect on the project, explained to me that the best shape for the roof of the 1.2km-long building was a shell with a catenary cross-section. Even though the catenary would be gigantic – 45m wide at the base and 39m high in the middle – the efficiency with which it would distribute its own weight meant that it only needed to be 16cm thick. 'A non-catenary curve might be perfectly doable, but it takes more material, it has bigger beam sections, and overall it is much more complicated to construct,' he said. 'Even if the cladding falls out, the interiors and everything else falls away and the whole thing turns to dust and rubble and sand, [the catenary] should still stand.'

The office of Foster and Partners contains detailed models of its most famous projects: London's 'Gherkin' tower, the Reichstag in Berlin and the suspension bridge at Millau, France. Yet on the table in front of Nikolai Malsch was a hanging bicycle chain. 'We love the catenary,' he said, 'because it tells the story of holding up the roof.'

The catenary has another, less widely known role in architecture, unlikely to feature in the design of churches and airports. A track of upside-down catenary humps is the required surface for the smooth riding of a square-wheeled bicycle, or for a bowling alley with cube-shaped balls.

Mathematician Stan Wagon rides his trike at Macalester College, St Paul, Minnesota.

Even though the Bernoullis produced more significant mathematicians than any other family in history, the greatest mathematician to emerge from Basel during their lifetimes wasn't one of them. Leonhard Euler (pronounced OIL-er) was the precociously clever son of a local pastor, with mathematical gifts that were first discovered, and then nurtured, by Johann Bernoulli, his Sunday-afternoon tutor. When he was 19, in 1727, Euler emigrated to Russia to take up a position at the newly opened Saint Petersburg Academy of Sciences, where Johann's son Daniel held the chair of mathematics. Peter the Great was offering royal salaries to lure over the brightest European minds, and Saint Petersburg provided a much more stimulating and high-profile intellectual environment than Basel. Euler was soon its most celebrated scholar.

Euler was a tranquil, devoted family man quite unlike the cliché of the socially awkward mathematical genius. His quirks were a prodigious memory – it is said he could recall all ten thousand lines of Virgil's *Aeneid* – and an even more prodigious work rate. No mathematician has come close to equalling his output, which averaged about 800 pages a year. When he died aged 76, in 1783, so much was left on his desk that his articles continued to appear in journals for another half century. Euler struggled with poor sight all his life, losing the use of his right eye in his late twenties and his left in his mid-sixties. He dictated some of his most important work while blind, to a roomful of scribes who scrambled to keep up. He could create mathematics, they said, faster than you could write it down.

But it wasn't just the quantity of his research that makes Euler stand out. It was also the quality and the diversity. 'Read Euler, read Euler,' implored the French mathematician Pierre-Simon Laplace. 'He is the master of us all.' Euler made important contributions to almost every field at the time, from number theory to mechanics, and from geometry to probability, as well as inventing new ones, too. His work was so transformational that the maths community adopted his symbolic vocabulary. Euler is the reason, for example, that we use π and e for the circle and exponential constants. He was not the first person to write π – that was a little-known Welsh mathematician, William Jones – but it was only because Euler employed the symbol that it gained widespread use. He was, however, the first person to use the letter e for the exponential constant, in a manuscript about the ballistics of cannonballs. The assumption is that he chose e because it was the first 'unused' letter available – mathematical texts were already full of a's, b's, c's and d's – rather than naming it after the word 'exponential' or his own surname. Despite his successes, he remained a modest man.

Euler made an unexpected discovery about e, which I will come to once I introduce a new symbol, the exclamation mark (which was not one of Euler's coinages). When a '!' is written immediately after a whole number it means that the number must be multiplied by all the whole numbers smaller than it. The '!' operation is called the 'factorial', so the number $n!$ is read 'n-factorial'.

The factorials begin:

(0! = 1 by convention)
1! = 1
2! = 2 × 1 = 2
3! = 3 × 2 × 1 = 6
4! = 4 × 3 × 2 × 1 = 24
...
10! = 10 × 9 × 8 × 7 × 6 × 5 × 4 × 3 × 2 × 1 = 3,628,800
...

Factorials grow fast. By the time we get to 20! the value is in the quintillions. The decision by German mathematicians in the nineteenth century to adopt the exclamation mark was possibly a comment on this phenomenal acceleration. Some English texts from that time even suggested that *n*! should be read '*n*-admiration' rather than '*n*-factorial'. Certainly, one can only admire the upwards trajectory of the exclamation mark: factorial outpaces even exponential growth.

Factorials appear most frequently when calculating combinations and permutations. For example, how many ways can you sit a certain number of people on the same number of chairs? Trivially, a single person can sit on a single chair in just one way. With two people and two chairs, there are two choices, the permutations AB and BA. With three people and three chairs there are six ways: ABC, ACB, BAC, BCA, CAB and CBA. Rather than listing all the permutations, however, there is a general method to find the total. The first sitter has three choices of chair, the second sitter has two and the third has one, so the total is 3 × 2 × 1 = 6. Adopting the same method with four people and four chairs, we find that the total is 4 × 3 × 2 × 1 = 4! = 24. In other words, if you have *n* people and *n* chairs the number of permutations is *n*! It is striking to realize that if you have a dinner party for ten people, you can fit them around a table in more than three and a half million ways.

Now let's return to *e*. The exponential constant can be written with a whole bunch of exclamation marks. OMG!!! LOL!!! It turns out that if we calculate $\frac{1}{n!}$ for every number starting from 0, then add up all the terms, the answer is e.

Written as an equation:

$$e = \frac{1}{0!} + \frac{1}{1!} + \frac{1}{2!} + \frac{1}{3!} + \frac{1}{4!} + \frac{1}{5!} + \cdots$$

Which is:

$$e = 1 + 1 + \frac{1}{2} + \frac{1}{6} + \frac{1}{24} + \frac{1}{120} + \cdots$$

Let's start counting term by term.

1
2
2.5
2.6666...
2.7083...
2.7166...

The series bears down on the true value of e with supersonic speed. After only ten terms it is accurate to six decimal places, which is good enough for almost any scientific application.

Why is e so beautifully expressed using factorials? As we saw with compound interest, the number is the limit of $(1 + \frac{1}{n})^n$ as n approaches infinity. I will spare you the details of the proof, but the term $(1 + \frac{1}{n})^n$ can be expanded into a giant sum that reduces into unit fractions with factorials in their denominators.

Euler's approach to research was playful, and he often investigated frivolous games and puzzles. When a chess enthusiast asked him, for example, whether it was possible for a knight to move across the board so it would land on every square only once before returning to its point of departure, Euler discovered how to do it, presaging similar questions still researched today. Euler also became intrigued by the French card game *jeu de rencontre*, or the game of coincidence, a variation of one of my favourite pastimes as a child, Snap!

In *rencontre*, two players A and B have a shuffled deck of cards each. They both turn over a card from their respective packs at the same time, and continue revealing cards simultaneously until all are used up. If at any turn the two cards are identical, then A wins. (And I shout 'Snap!') If they get through the pack with no matches, then B wins. Euler wanted to know the likelihood of A winning the game; that is, of there being at least one coincidence in 52 turns.

This question has reappeared in many guises over the years. Imagine, for example, that a cloakroom attendant fails to tag any of the items that are handed in over an evening, and gives them back randomly to guests at the end of the night. What is the chance that at least one person receives the right coat back? Or say that a cinema sells numbered tickets but then lets members of the audience sit wherever they like. If the theatre is full, what is the chance that at least one seat is occupied by a person with the correct ticket?

Euler started at the beginning. When the *rencontre* pack given to each player consists of one card, there is a 100 per cent chance of a match. When the pack consists of two cards, there is a 50 per cent chance. Euler drew permutation tables for games played with three-card and four-card packs before deducing the pattern. The probability of a match when there are n cards in the pack is the fraction:

$$1 - \frac{1}{2!} + \frac{1}{3!} - \frac{1}{4!} - \dots \pm \frac{1}{n!}$$

Hang on! This pattern looks similar to the series for e above.

I'll skip the guts of the proof, but this series approximates to $(1 - \frac{1}{e})$, or about 0.63. The series is only exactly $(1 - \frac{1}{e})$ when n gets to infinity, but the approximation is already very, very good after a few terms. When $n = 52$, the number of cards in a pack, the series is $(1 - \frac{1}{e})$ correct to almost 70 decimal places.

The chance of a coincidence in *jeu de rencontre*, therefore, is about 63 per cent. It's roughly twice as likely to happen as it is not. Likewise, the chance of at least one party guest getting his coat back is 63 per cent, and the chance of a cinema-goer sitting on the correct seat is also 63 per cent. It is interesting to note that the number of cards in the deck, guests handing in coats, or seats in the cinema makes little difference to the chances of at least one match, provided there are more than six or seven cards, guests or seats. Each time you increase the number of cards, guests or seats, you add an extra term in the series above that determines the probability of a match. An eighth card, for example, gives you an eighth term, $\frac{1}{8!}$, or 0.0000248, which alters the probability by less than a quarter of a hundredth of one per cent. A ninth card alters the probability by even less. In other words, the chances of a match barely change whether you're playing with a full deck or just the cards of one suit. Similarly, it makes almost no difference whether ten or a hundred guests check their coats, or if the cinema is the smallest screen in your local multiplex or the Empire Leicester Square.

Euler's discovery that *e* is embedded in the maths of card games is one of the earliest instances of the constant appearing in an area with no obvious connection to exponential growth. It would later appear in an equally competitive arena: the maths of finding a wife.

Let's briefly return to Johannes Kepler. After the German astronomer was widowed in 1611, he interviewed eleven women for the position of the second Frau K. The task started badly, he wrote: the first candidate had 'stinking breath', the second 'had been brought up in luxury that was above her station', and the third was engaged to a man who had sired a child with a prostitute. He would have married the fourth, of 'tall stature and athletic build' had he not seen the fifth, who promised to be 'modest, thrifty, diligent and to love her step-children'. But his prevaricating caused both women to lose interest and he saw a sixth, whom he ruled out because he 'feared the expense of a sumptuous wedding', and a seventh, who, despite her 'appearance which deserved to be loved', rejected him when he again did not decide quickly enough. The eighth 'had nothing to recommend her, [even though] her mother was a most worthy person', the ninth had lung disease, the tenth had a 'shape ugly even for a man of simple tastes ... short and fat, and coming from a family distinguished by redundant obesity', and the final candidate was not grown-up enough. At the end of the process Kepler asked: 'Was it Divine Providence or my own moral guilt which, for two years or longer, tore me in so many different directions and made me consider the possibilities of such different unions?' Such agonized self-searching is familiar in the modern dating scene. What the great German astronomer needed was a strategy.

Consider the following game, which, according to the maths author Martin Gardner, was invented by a couple of friends, John H. Fox and L. Gerald Marnie, in 1958. Ask someone to take as many pieces of paper as they like, and to write a different positive number on each one. The numbers can be anything from a tiny fraction to something absurdly enormous, say 1 with a hundred zeros after it. The pieces of paper are placed face down on a table and shuffled around. Now the game begins. You turn over the pieces of paper one by one. The aim is to stop when you turn over the paper with the largest number on it. You are not allowed to go back and pick a number on a paper that you have already turned. And if you keep on going and turn over all the papers, your pick is the last one you turned.

Since the player turning over the papers has no idea of the numbers written on them, you might think that his or her chances of picking the highest number are slim. Astoundingly, however, it is possible to win this game more than a third of the time, irrespective of how many pieces of paper are being used. The trick is to use information from numbers you have already seen to make a judgement about the numbers that remain face down. The strategy: turn over a certain quantity of the slips, benchmark the highest number among this selection, and then pick the first number you turn over that is higher than the benchmark. The optimal solution, in fact, is to turn over $\frac{1}{e}$ (or 0.368, or 36.8 per cent) of the total number of pieces of paper and then to choose the first number higher than any number in that selection. If you do, the chance of picking the highest number is again $\frac{1}{e}$, or 36.8 per cent.

In the 1960s this conundrum became known as the Secretary Problem, or the Marriage Problem, as it was analogous to the situation of a boss looking at a list of applicants, or a man looking at a list of potential wives, and having to decide how to choose the best candidate. (And evidently because mathematicians were mainly men.)

Imagine you are interviewing twenty people to be your secretary, with the rule that you must decide at the end of each interview whether or not to give that applicant the job. If you offer the job to the first candidate, you cannot see any of the others, and if you haven't chosen anyone by the time you see the last candidate you must offer the job to her. Or imagine you have decided to date twenty women, with the proviso that you must decide whether she is The One before moving on to the next. (Apologies to female readers. The analogy is based on the notion that it's men who propose to women, and that, once proposed to, a woman will always say yes.) If you propose on the first date, you are not allowed to meet any of the others, but if you date them all you must propose to the final one you see. In both cases the way to maximise your chances of choosing the best match is to interview or date 36.8 per cent of the candidates and then offer the job or propose to the next candidate better than all those who came before. The method will not *guarantee* that you find the best match – the chances are only 36.8 per cent – but it is still the best strategy overall.

If Kepler had realized he was going to interview eleven women, and had followed this strategy, he would have interviewed 36.8 per cent of them, which works out as four, and then proposed to the next one he liked more than any of those. In other words, he would have chosen the fifth, which is what he did indeed do once he had seen all eleven (it turned out to be a happy marriage). Had he known the Marriage Problem and its solution, Kepler would have saved himself six bad dates.

The Secretary/Marriage Problem has become one of the most famous questions in recreational mathematics, even though it does not reflect reality, since bosses can recall candidates and men can return (as Kepler did) to previous dates. Underlying the whimsy, however, is a whole field of incredibly useful theory, called 'optimal stopping', or the maths of when is the best time to stop. Optimal stopping is important in finance, for working out when it's time to cut one's losses on an investment, or when to exercise a stock option. But it also comes into areas as diverse as medicine (to, say, calculate the best time to stop a particular treatment); economics (to project when is the best time to stop relying on fossil fuels, for example); and zoology (deciding, say, when is the best time to stop searching a large population of animals for new species, and so avoid wasting money looking for what probably isn't there).

Boris Berezovsky, the billionaire Russian oligarch, used to be a maths professor at the USSR Academy of Sciences, the Soviet descendent of Euler's alma mater. While there in the 1980s, he co-authored a book on the Secretary Problem. He moved to the United Kingdom in 2003. I approached him for an interview several times, but each time we spoke he asked me to call back in a couple of months. After a year of trying, I figured it was optimal to stop.

The point about optimal stopping is that it is possible to make informed decisions about random events based on accumulated knowledge. Here is a game with a fantastically ingenious way of making use of the tiniest amount of information (it does not concern *e*, but please forgive me for this brief detour from exponentials). The result is so bafflingly counterintuitive that when it first did the rounds, many mathematicians did not believe it.

The game is simple. You write down two different numbers on separate pieces of paper and then place them both face down. I will turn over one of the numbers and tell you whether or not it is bigger than the one that remains hidden. Amazing though it seems, I will get the answer right more than half the time.

My performance sounds like magic but there is no trick. Nor does my strategy rely on the human element of how you selected the numbers, or how you placed the paper on the table. Mathematics, not psychology, provides me with a method to win more times than I lose.

Just say that I am not allowed to turn over any of the pieces of paper. The odds of guessing which one has the highest number on it are 50/50. There are two choices, and one will be correct. My chances of getting the right answer are the same as flipping a coin.

But when I see one of the numbers, however, my odds will improve if I take the following steps:

(1) I generate a random number myself – let's call it *k*.

(2) If *k* is less than the number I turned over, I say the number I turned over is highest.

(3) If *k* is more than the number I turned over, I say the number that remains face down is highest.

My strategy, in other words, is to go with the number I see, unless my random number *k* is larger. In this case I switch to the one I have not seen.

To see why the strategy gives me an edge, we need to consider the value of *k* with respect to the two numbers on the pieces of paper. There are three possibilities: (i) *k* is less than both of them, (ii) *k* is higher than both of them, or (iii) *k* is between them.

In the first scenario, whatever number I see, I stick with it. My chances of being right are 50/50. In the second scenario, whatever number I see, I choose the other one. My chances are again 50/50. The interesting situation is the third one, when I win 100 per cent of the time. If I see the lowest number, I switch, and if I see the highest number, I stick with it. When my random number serendipitously falls in the middle of the two numbers you wrote on the pieces of paper, I will always win!

(I need to explain in a little more detail how I generate *k*, since for the strategy to work *k* must always have a chance of being between any two given numbers. Otherwise, it is easy to find scenarios where I do not have an edge. For example, if you always write down negative numbers and my random number is always positive then my number will never be between yours and my odds of success stay at 50/50. My solution is to choose a random number from a 'normal distribution', since this gives a chance for all positive and negative numbers to come up. You don't need to know anything more about normal distributions other than that they provide a way of giving a random number that has a chance of being between any two other numbers.)

There may only be a small chance that *k* will fall between your numbers. But because there is a chance, no matter how meagre, it means that if we play this game enough times my overall odds of winning increase beyond 50 per cent. I will never know a priori when I will win and when I will lose. But I never promised that. All I said is that I can win more than half the time. If you want to make sure that my odds stay as close to 50/50 as possible, you need to choose two numbers that are as close together as possible. Yet as long as they are not equal, there is always the chance that I will choose a number between them, and as long as this chance is mathematically possible, I will win the game more frequently than I will lose it.

Last chapter I introduced pi, which begins 3.14159 and is the number of times the diameter fits round the circumference of a circle. This chapter we got to know *e*, which begins 2.71828 and is the numerical essence of exponential growth. The numbers are the two most frequently used mathematical constants, and are often spoken of together, even though they emerged from different quests and have different mathematical personalities. It is curious that they are close to each other, less than 0.5 apart. In 1859 the American mathematician Benjamin Peirce introduced the symbol ⋔ for pi and the symbol ⋔ for *e*, as if to show that they are both somehow images of each other, but his confusing notation never caught on.

Both constants are irrational numbers, which means that the digits in their decimal expansions go on forever without endlessly repeating themselves. It has become something of a mathematical sport to try to find arithmetical combinations of both terms that are as elegant as possible. We are never going to find equations with absolute equality, but:

$$\pi^4 + \pi^5 = e^6$$
which is correct to seven significant figures.

In a similar vein:
$$e^\pi - \pi = 19.999099979\ldots$$
which is very, very close to 20.

And, most impressively:
$$e^{\pi\sqrt{163}} = 262537412640768743.99999999999925007\ldots$$
which is less than a trillionth from a whole number!

166

In 1730 the Scottish mathematician James Stirling discovered the following formula:

$$n! \approx \sqrt{(2\pi n)} n^n e^{-n}$$

It gives an approximation for the value of $n!$, the factorial of n, which as we saw above is the number you get when you multiply $1 \times 2 \times 3 \times 4 \times \dots \times n$.

The factorial is a simple procedure, just multiplying whole numbers together, so it is arresting to see that on the right side of the formula there is a square root sign, a pi and an *e*.

When $n = 10$, the approximation is less than one per cent off the real value of 10!, and the bigger n gets the more accurate the approximation becomes in percentage terms. Since factorials are such huge numbers (10! is 3,628,800), the formula is gobsmacking.

Something's going on between pi and *e*.

Leonhard Euler uncovered another coupling between the two constants, one even more surprising and stunning than Stirling's formula. But before we get there we need to familiarize ourselves with another vowel that he introduced to our mathematical alphabet.

Prepare to meet *i*.

The Positive Power of Negative Thinking

In the winter of 2007 the UK National Lottery launched a new scratchcard. Two numbers were on each card and you won a prize if the number on the left was higher than the number on the right. Pretty straightforward, you might think. Yet because the card was winter-themed, the numbers were temperatures below freezing. The task, therefore, was to compare *negative* numbers, and for some this was not straightforward at all. Many players were unable to grasp that, for example, –8 is lower than –6, and after dozens of complaints the card was taken off the market. 'They fobbed me off with some story that –6 is higher, not lower, than –8, but I'm not having it,' one disgruntled punter protested.

It's easy to poke fun at those who cannot understand basic arithmetic, yet we should not chuckle too heartily. Negative numbers have caused us centuries of mental torment. They still do. That's why it's much more common for underground floors in buildings to be marked with letters such as LG and B, or alphanumeric combinations such as B1, B2 and B3, than –1, –2 and –3. When we date events that happened before the birth of Christ, like the year Euclid wrote *Elements*, we prefer to say around 300 BCE rather than around –300 CE. And accountants have myriad ways of avoiding the minus sign: writing debts in red, adding the abbreviation DR (for debtor) or smuggling the disagreeable sum away in brackets.

Neither Greek nor Egyptian nor Babylonian mathematics developed the concept of negative numbers. For the ancients, numbers were for counting and measuring, and how can you count or measure less than nothing of something? Let's put ourselves inside their minds to understand the mental leap required. We know that 2 + 3 = 5 because when we have two loaves of bread and are given three more we have five loaves. And we know that 2 – 1 = 1 because when we have two loaves and give one away, we still have one left. Yet what can 2 – 3 mean? If I have only two loaves I cannot possibly give three away. Let's say, however, that I *can* do this, leaving me with minus one loaf. What is this 'minus one loaf'? It is not a type of loaf. Rather it is the absence of a loaf, such that when a loaf is added to it there is nothing there. No wonder the ancients found the notion absurd.

In ancient Asia, however, they did allow negative quantities – up to a point. By the time of Euclid, the Chinese had a system of calculating that used arrangements of bamboo rods. Normal rods described positive numbers, which they called 'true', and rods painted black described negative numbers, which they called 'false'. They placed the rods on a checkerboard, as illustrated below, such that each number was on an individual square and each column represented an equation. A skilled calculator could solve the equations by moving the rods around. If the solution was made up of normal rods, it was a true number, and it was accepted. If the solution was made up of black rods, it was a false number, and it was discarded. That the Chinese used physical objects to represent negative quantities showed that these numbers did exist in some way, even if they were just tools for calculating positive amounts. The Chinese had grasped an essential truth: if mathematical objects are useful, who cares if they do not correspond with everyday experience? Leave that problem to the philosophers.

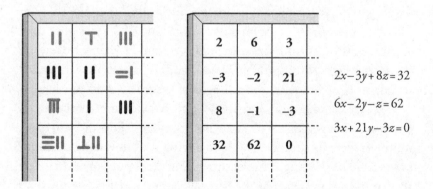

The Chinese placed rods on a checkerboard – normal ones for positive numbers, black ones for negative numbers – that spelled out equations.

A few centuries later, in India, mathematicians found a tangible context for negative quantities: money. If I borrow five rupees from you, I have a debt of five rupees, a negative quantity that will only become zero when I give you five rupees back. The seventh-century astronomer Brahmagupta composed rules for the arithmetic of positive and negative numbers, which he called 'fortunes' and 'debts'. It was the same text in which he introduced the modern number zero.

> *A debt minus zero is a debt.*
> *A fortune minus zero is a fortune.*
> *Zero minus zero is zero.*
> *A debt subtracted from zero is a fortune.*
> *A fortune subtracted from zero is a debt.*
> And so on …

Brahmagupta described the precise value of fortunes and debts using zero and nine other digits, which as we saw earlier is the origin of the decimal notation we use today. Indian numerals spread to the Middle East, north Africa and, by the tenth century, Spain. Even so, it took three more centuries before negative numbers were widely used in Europe. This delay was down to several factors: because of the historic connection with debts, and therefore the sinful practice of usury; because of a general suspicion about new-fangled methods arriving from Muslim lands; and because of the enduring influence of Greek thought, in which you cannot have less of a quantity than none of it.

Accountants eventually became comfortable with using negative numbers in their professional lives, but for a long time mathematicians remained wary of them. In the fifteenth and sixteenth centuries the negatives were known as absurd numbers – *numeri absurdi* – and even in the seventeenth many regarded them as nonsensical. The following argument against negative numbers persisted into the eighteenth century. Consider this equation:

$$\frac{-1}{1} = \frac{1}{-1}$$

The statement is arithmetically true, yet it is also paradoxical, since it states that the ratio of a smaller number, −1, to a larger number, 1, is equal to the ratio of a larger number, 1, to a smaller one, −1. The paradox was much discussed, and no one could untangle it. To make sense of negative numbers, many mathematicians, including Leonhard Euler, came to the bizarre conclusion that they were larger than infinity. This notion comes from considering the following sequence:

$$\frac{10}{3}, \frac{10}{2}, \frac{10}{1}, \frac{10}{\frac{1}{2}} \dots$$

Which is:

3.3, 5, 10, 20 …

As the number on the bottom of the fraction, the denominator, gets lower, from 3 to 2 to 1 to $\frac{1}{2}$, the absolute value of the fraction gets higher, and as the denominator approaches zero, the absolute value of the fraction approaches infinity. It was assumed that when the denominator reaches zero, the fraction is infinity, and when the denominator is less than zero – that is, when it is negative – the fraction must be larger than infinity. Nowadays, we avoid this conundrum by claiming that it is meaningless to divide a number by zero. The fraction $\frac{10}{0}$ is not infinity, but 'undefined'.

One voice of clarity among the confusion belonged to the English mathematician John Wallis, who devised a powerful visual interpretation for the negative numbers. In his 1685 work *A Treatise of Algebra,* he first described the 'number line', illustrated opposite, in which positive and negative numbers represent distances from zero in opposite directions. Wallis wrote that a man walking five yards forwards from zero and then eight yards backwards 'is advanced 3 Yards less than nothing … And consequently −3, doth as truly design [a point on the line]; as +3 designed [a point on the line]. Not Forward, as was supposed; but Backward.' By replacing the idea of quantity with the idea of position, Wallis argued that negative numbers were neither 'Unuseful [nor] Absurd', which turned out to be something of an understatement. It took a few years for Wallis's idea to enter the mainstream, but in retrospect it is the most successful explicatory diagram of all time. It has endless practical applications, from graphs to thermometers. We have no conceptual difficulties in imagining negative numbers now we can see them on a line.

The number line.

The German philosopher Immanuel Kant entered the debate about negative numbers when he declared, in his *Attempt to Introduce the Concept of Negative Quantities into World-Wisdom*, that it was pointless using metaphysical arguments against them. He showed that in the real world many things can have positive and negative values, such as two opposing forces acting on an object. A negative number does not represent the denial of a number, but rather a compatible opposite.

Yet even at the end of the eighteenth century, some mathematicians held on to the belief that negative numbers were 'a jargon, at which common sense recoils; but, from its having been once adopted, like many other figments, it finds the most strenuous supporters among those who love to take things upon trust, and hate the labour of a serious thought.' William Frend, who was the second-best maths student in his year at Cambridge, wrote these words in 1796, in a book which is unique in mathematical literature: an introduction to algebra containing not a single negative number.

When we learn negative numbers at school we are not exposed to these past controversies. We accept the negatives by analogy to the line, and are then presented with a bombshell:

Minus times minus equals plus.

Gulp! The number line is wonderful at providing a visual representation of negative numbers, but it gives us no insight into what happens when we multiply them by each other. Maths just got harder.

Why does the multiplication of two negatives equal a positive? Because it follows from the rules of multiplication that work with positive numbers. We accept that two negatives make a positive because it keeps arithmetic coherent, not because it has any meaning beyond the system. It is a necessary piece of structural underpinning that makes sure the house of numbers does not collapse. Consider the number line. If I walk two steps forward from 0, I reach 2. If I repeat these two steps I reach 4, and if I repeat them again I get to 6. Likewise if I walk two units back from zero I reach –2, and if I repeat the steps twice more I get to –6. We can interpret these procedures as the equations:

$$2 + 2 + 2 = 6$$
$$-2 - 2 - 2 = -6$$

Which are equivalent to the multiplications:

$$3 \times 2 = 6$$
$$3 \times -2 = -6$$

These equations tell us that a positive times a positive equals a positive, and a positive times a negative equals a negative. To find out what happens to the multiplication of two negatives, let's substitute 3 with $(4 - 1)$ in the last equation to get:

$$(4 - 1) \times -2 = -6$$

We can rewrite this equation as

$$(4 \times -2) + (-1 \times -2) = -6$$

since we know from the arithmetic of positive numbers that when you multiply two terms in a bracket by a single number, you must multiply each number in the bracket individually. (This rule is known as the 'distributive law'.) The equation becomes:

$$-8 + (-1 \times -2) = -6$$

So:

$$(-1 \times -2) = 2$$

And there we have it. *Minus times minus equals plus.*

One reason we find the multiplication of negatives so conceptually troubling is that there are many situations in life in which arithmetic provides the wrong model. No sooner has teacher explained the idea to us than we are lectured that two wrongs don't make a right. In linguistics, double negatives can be either negatives or positives, depending on the context and the language spoken. When I learnt Portuguese, I had to get used to the idea that 'I know nothing' is *não sei nada,* or 'I don't know nothing'. In this case the two negatives reinforce the negative, rather than cancelling each other out.

A double negative in English, of course, is a positive. The linguist J. L. Austin once told a conference that there are no languages in which two positives make a negative. It is said that the philosopher Sidney Morgenbesser, sitting in the audience, replied: 'Yeah, *yeah.*'

An early champion of the Indian number system with a zero and negatives was the Arab mathematician Muhammad ibn Musa al-Khwarizmi (c.750–c.850). Latin versions of his surname were later used to describe the arithmetical techniques he publicized, and are the root of the word 'algorithm'. Al-Khwarizmi also developed a new type of mathematics, algebra, the name of which comes from the Arabic word *al-jabr,* meaning restoration. Algebra is the language of equations, in which symbols like x and y are used to represent numbers. In algebra, 'What number added to two equals zero?' can be rephrased as the quest to find x when:

$$x + 2 = 0$$

The answer is $x = -2$. Whether you believed negative numbers were meaningful or not, the term -2 is the solution to the equation. It was thanks to algebra that the European mathematicians of the Renaissance finally expanded the definition of 'number' to include the negatives. Absurd they may have been, but they were numbers all the same.

Soon algebraists ran into another problem. Using only the positive and negative numbers and the four arithmetical operations of addition, subtraction, multiplication and division, they stumbled on a concept they could not understand. It was the solution to the equation:

$$x^2 = -1$$

The answer is 'the square root of minus one', or $x = \sqrt{-1}$. But here's the headache: 'What sort of number, when multiplied by itself, is a negative?' It cannot be positive, since a positive times a positive is a positive. Nor can it be negative, since a negative times a negative is also a positive. The first person to consider the square root of a negative number was the Italian mathematician Girolamo Cardano, in 1545. He said thinking about it caused him 'mental tortures', as it will anyone who has not encountered the concept before. So he ignored it, declaring that if the solution to an equation is the square root of a negative number then it is 'as refined as it is useless'. Cardano had opened the door on a whole new world of maths, and then slammed it shut again.

A few decades later, Cardano's compatriot Rafael Bombelli reopened the door and timorously walked through it. The square roots of negative numbers were appearing more and more in algebraic calculations, so Bombelli decided to treat them just like positive and negative numbers, adding them, subtracting them, multiplying and dividing them as and when they cropped up. 'It was a wild thought in the judgement of many,' he wrote. 'The whole matter seemed to rest on sophistry rather than on truth.' Yet not only were the square roots of negative numbers well behaved, they also allowed him to solve equations that had not been solvable before. If you didn't worry about what they meant, they could be accommodated within the fold.

In 1637 René Descartes described the square roots of negative numbers as 'imaginary', a word given the stamp of approval by Leonhard Euler a century later: 'All such expressions as $\sqrt{-1}$, $\sqrt{-2}$, etc., are consequently impossible or imaginary numbers, since they represent roots of negative quantities; and of such numbers we may truly assert that they are neither nothing, nor greater than nothing, nor less than nothing, which necessarily constitutes them imaginary or impossible.' Euler gave the number $\sqrt{-1}$ its own symbol, i, for 'imaginary', and showed that the square root of every negative number can be expressed as a multiple of i. For example, $\sqrt{-4}$ becomes $2i$, since $\sqrt{-4} = \sqrt{4 \times -1} = \sqrt{4} \times \sqrt{-1} = 2 \times i = 2i$. More generally, $\sqrt{-n} = (\sqrt{n})i$. The square roots of negative numbers – which are all multiples of i – are collectively known as the 'imaginary numbers'.

Numbers that are not imaginary are known as the 'real' numbers. They are real because they sit on the number line, so we can see that they are really there. The numbers 2, 3.5, −4 and π are all real, but the numbers $2i$, $3.5i$, $-4i$ and πi are all imaginary. In fact, the imaginary numbers are a kind of mirror family to the real numbers. For every real number m there is an imaginary number mi.

When a real number is added to an imaginary number, the hybrid form, such as $3 + 2i$, is called a 'complex number'. All complex numbers are of the form $a + bi$, where a and b are real numbers and i is $\sqrt{-1}$. Since you can't add a real number to an imaginary number in the traditional sense, the plus sign is just a way of separating the two parts. A complex number is considered a single number with two parts, its real and imaginary ones. If the real part is zero, the number is purely imaginary, and if the imaginary part is zero, the number is purely real.

The meaning of 'number', a concept first used to count physical things, was extended by the introduction of negatives, and it was extended again by the introduction of imaginaries. Inevitably the question arose as to whether algebra would lead to another even more abstract type of number. What is the square root of the square root of minus one, for example? A concept that, if you think about it too hard, will turn your brain upside down and twist it inside out. It is the solution to the equation:

$$x = \sqrt{(\sqrt{-1})}$$

Or:

$$x^2 = \sqrt{-1}$$

Which is:

$$x^2 = i$$

Amazingly, the solution is $x = \frac{1}{\sqrt{2}} + (\frac{1}{\sqrt{2}})i$, a complex number.*

In the eighteenth century, mathematicians realized that to solve equations no numbers are needed beyond the imaginaries, a result so important it is known as the Fundamental Theorem of Algebra. Every equation written with complex numbers will always produce a solution with complex numbers. The door that Rafael Bombelli walked through to investigate the square roots of negative numbers revealed a solitary room. But what a room it was! The squeamishness mathematicians once had with imaginary numbers has been replaced by joy. The concept of i is now considered a very natural and efficient extension of the number system. For the price of a solitary symbol, mathematicians gained an elegantly self-contained abstract universe. Bargain.

Imaginary numbers are protagonists of two of the most famous examples of mathematical beauty. One is a picture (of which more later) and one is an equation, known as Euler's identity, which in 2003 was sprayed on the side of an SUV in an eco-terrorist attack on a Los Angeles car dealership. The nature of the graffiti led to the arrest of a physics PhD student at Caltech. 'Everyone should know Euler's [identity],' he explained to the judge. He was correct, but one should nevertheless refrain from daubing it on cars. Euler's identity is the 'To be or not to be' of mathematics, the most famous line in the oeuvre and a piece of cultural heritage that resonates beyond its field:

$$e^{i\pi} + 1 = 0$$

---

* When we multiply complex numbers we obey the normal rules of arithmetic. I won't prove it here, but we can assume that for any numbers a and b, either real or imaginary, then $(a + b)^2 = a^2 + 2ab + b^2$. So, if $x = \frac{1}{\sqrt{2}} + (\frac{1}{\sqrt{2}})i$, then $x^2 = (\frac{1}{\sqrt{2}} + (\frac{1}{\sqrt{2}})i)^2 = (\frac{1}{\sqrt{2}})^2 + (2 \times \frac{1}{\sqrt{2}} \times \frac{1}{\sqrt{2}}i) + (\frac{1}{\sqrt{2}}i)^2 = \frac{1}{2} + i + (-\frac{1}{2}) = i.$

The equation is mind-blowing. It cleanly unites the five most important numbers in maths: 1, the first counting number; 0, the abstraction of nothing; π, the ratio of a circle's circumference to its diameter; e, the exponential constant; and i, the square root of minus one. Each number emerged from a different area of enquiry, yet they unite with perfection. You couldn't have predicted a more immaculate synthesis of mathematical thought. Beauty in mathematics is about elegance of expression and making unexpected connections. No other equation is as concise or as deep.

What does it mean, however, for a real number, e, to have an imaginary power, $i\pi$? 'We cannot understand it, and we don't know what it means,' replied the nineteenth-century Harvard maths professor Benjamin Peirce. 'But we have proved it, and therefore we know it must be the truth.' Peirce was being arch. Mathematics starts from a set of assumptions and leads to wherever it goes. That's the fun of the ride. Indeed, it was by forgetting about meaning that Euler made his discovery. Since his identity is the most celebrated equation in mathematics I would be doing you a disservice if I did not at least outline the story.

Our only prep will be to assume, without proof, the following three equations. The dots mean that the right-hand side of the equation continues for an infinite number of terms:

$$e^x = 1 + x + \frac{x^2}{2!} + \frac{x^3}{3!} + \frac{x^4}{4!} + \frac{x^5}{5!} + \cdots$$

$$\sin x = x - \frac{x^3}{3!} + \frac{x^5}{5!} - \frac{x^7}{7!} + \frac{x^9}{9!} - \frac{x^{11}}{11!} + \cdots$$

$$\cos x = 1 - \frac{x^2}{2!} + \frac{x^4}{4!} - \frac{x^6}{6!} + \frac{x^8}{8!} - \frac{x^{10}}{10!} + \cdots$$

If x is equal to 1 then the first series gives us the formula for calculating the exponential constant e, which we encountered in the previous chapter. (Remember that the factorial of a number n, written $n!$, means that the number is multiplied by all the numbers from 1 to n). The next two infinite series are for the sine and cosine of x, the trigonometric ratios which should also be familiar from earlier. For the sine and cosine series to work, however, we must use a bespoke unit of measurement, the radian, and not the traditional unit of measurement, the degree. A full circle, or 360 degrees, is 2π radians, and half a circle, 180 degrees, is π radians. (A radian is so called because 1 radian is the angle at the centre of a circle that subtends an arc of the circumference equal to the radius, as shown below. The radian is a much more natural way to define an angle than the Babylonian-inspired degree system we discussed in an earlier chapter, and since the eighteenth century, mathematicians have preferred it.)

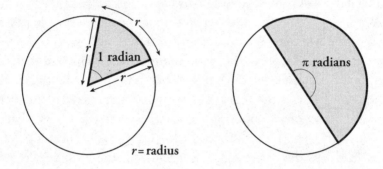

The radian.

While there is no intuitive way to understand what it is to raise a number like e to an imaginary power, Euler realized that we can at least do it algebraically using the infinite series for e^x on the previous page. If, for example, we substitute ix for x, we get the following equation:

$$e^{ix} = 1 + ix + \frac{(ix)^2}{2!} + \frac{(ix)^3}{3!} + \frac{(ix)^4}{4!} + \frac{(ix)^5}{5!} + \dots$$

Which, when the brackets are removed, becomes:

$$e^{ix} = 1 + ix + \frac{i^2x^2}{2!} + \frac{i^3x^3}{3!} + \frac{i^4x^4}{4!} + \frac{i^5x^5}{5!} + \dots$$

We can simplify this further, since, by definition, $i^2 = -1$, and:

$i^3 = i \times i \times i = i^2 \times i = -1 \times i = -i,$
$i^4 = i^2 \times i^2 = -1 \times -1 = 1,$
$i^5 = i^4 \times i = 1 \times i = i,$
$i^6 = -1$

And so on.

In other words, we can replace the terms i^2, i^4, i^6, i^8 ... with the values -1, 1, -1, 1 ... and we can replace the terms i^3, i^5, i^7, i^9 ... with the values $-i$, i, $-i$, i ... The equation can thus be rewritten:

$$e^{ix} = 1 + ix - \frac{x^2}{2!} - \frac{ix^3}{3!} + \frac{x^4}{4!} + \frac{ix^5}{5!} - \dots$$

The pattern is easier to see with the imaginary terms in bold:

$$e^{ix} = 1 + \boldsymbol{ix} - \frac{x^2}{2!} - \frac{\boldsymbol{ix^3}}{\boldsymbol{3!}} + \frac{x^4}{4!} + \frac{\boldsymbol{ix^5}}{\boldsymbol{5!}} - \dots$$

Which rearranges to:

$$e^{ix} = 1 - \frac{x^2}{2!} + \frac{x^4}{4!} - \frac{x^6}{6!} + \dots + i(\boldsymbol{x} - \frac{\boldsymbol{x^3}}{\boldsymbol{3!}} + \frac{\boldsymbol{x^5}}{\boldsymbol{5!}} - \frac{\boldsymbol{x^7}}{\boldsymbol{7!}} + \dots)$$

Ta dah! The terms are precisely those in the equations on p. 179 for the cosine and the sine of x:

$$e^{ix} = \cos x + i \sin x$$

By raising e to an imaginary power Euler found the trigonometric ratios. In other words, he took two familiar but unrelated concepts and stirred them together, and in a puff of smoke out popped something unexpected: two more familiar concepts from what was thought to have been a different field. It's mathematics, but it feels like alchemy.

To finish off, Euler said: Let $x = \pi$, which is the radian measure for 180°. Because $\cos \pi = \cos 180° = -1$, and $\sin \pi = \sin 180° = 0$, the imaginary term disappears.

$$e^{i\pi} = \cos \pi + i \sin \pi$$

Which reduces to:

$$e^{i\pi} = -1$$

Or:

$$e^{i\pi} + 1 = 0$$

Euler's groundbreaking work with imaginary numbers may have placed them at the centre of mathematics, where they have been ever since, but even so, to Euler and his eighteenth-century contemporaries the imaginaries remained exotic, mysterious beasts. A serious obstacle to embracing the imaginaries fully was their name, which implied that they did not exist. At the beginning of the eighteenth century Gottfried Leibniz had described $\sqrt{-1}$ as 'almost an amphibian between being and not being'. Maths may have advanced quicker had the term 'amphibious number' entered the vocabulary instead.

We saw earlier that mathematicians only became completely comfortable with negative numbers once they could see them on a page, representing points on a number line. History repeated itself with the imaginaries. Philosophical anxiety about complex numbers only disappeared with the invention of a simple visual interpretation.

The 'complex plane', illustrated below, consists of a horizontal number line for the real numbers, and a vertical number line for the imaginaries, just like the x and y axes in a coordinate graph. The complex number $a + bi$ is taken to mean the point on the complex plane with coordinates (a, b), meaning the point a along and b up. In the illustration I have marked the point $3 + 2i$, which is at point $(3, 2)$. The complex plane is not a complicated idea, and yet the three men who came up with it were all working independently

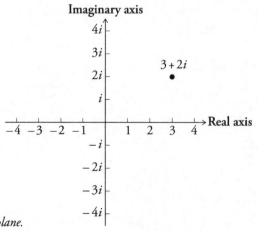

The complex plane.

on the fringes of the European mathematical establishment: Caspar Wessel, a surveyor in Copenhagen; Jean-Robert Argand, an accountant in Paris; and Abbé Adrien-Quentin Buée, a French cleric who had fled the Revolution and was living in Bath. The fact that none of the great mathematicians of the age conceived of the complex plane seems to reveal just how beholden they were to the doctrine that imaginary numbers exist only in the imagination.

The complex plane was a brilliant invention. Not only does it provide a map of where the complex numbers are, but it also enriches our understanding of how they behave.

Take a basic sum, say the addition of 1 to the number $3 + 2i$.
  The answer is $4 + 2i$.

Or let's add i to the number $3 + 2i$.
  The answer is $3 + 3i$.

Now look at the image below. Adding 1 to the point $3 + 2i$ moves us one unit along, and adding i moves us one unit up.
  The more 1s we add, the more we progress horizontally, and the more i's we add, the further we climb vertically. In fact, adding the complex number $a + bi$ is equivalent to moving a units along and b units up. We call this type of geometrical movement *translation*.

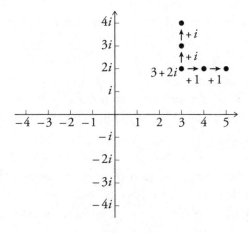

Now let's go forth and multiply. If we take $3 + 2i$ and multiply it by 1, we get the same number. Of course we do. That's what 1 always does. But when we multiply a number by i, something interesting occurs. Let's multiply $3 + 2i$ by i:

$$(3 + 2i) \times i = 3i + 2i^2 = 3i - 2 = -2 + 3i$$

Look at the image below. The point $3 + 2i$ has rotated 90 degrees anticlockwise around 0.

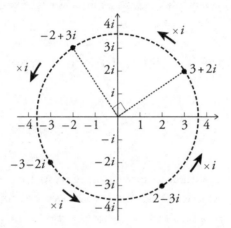

If we multiply this new point, $-2 + 3i$, by i, then again the result is a 90 degree rotation around 0.

In fact, when any complex number is multiplied by i, the point described by that number on the complex plane is rotated a quarter-turn around the origin. If we multiply by $i^2 = -1$, the point is rotated 180 degrees, if we multiply by $i^3 = -i$, the point is rotated 270 degrees, and if we multiply by $i^4 = 1$, the point is rotated back to the starting position.

Now take an arbitrary positive number a. It sits on the real axis of the complex plane. Multiply a by -1 and the answer is $-a$. This number also sits on the real axis, but it has flipped to the opposite position on the other side of 0. Multiply by -1 again and the number returns to a. If we multiply a by i, however, the answer is ai. The number has rotated 90 degrees and now sits on the imaginary axis. Multiply again by i, and the number rotates to position $-a$, back on the real axis. The complex plane thus lets us understand the *back and forth* multiplication of negative numbers as a consequence of the *round and round* multiplication of the imaginaries. Not only does this process give us a deeper sense of what numbers are, it also provides us with a powerful language for describing things that rotate.

Particle physics, electrical engineering and radar, among many other scientific fields, all rely on complex numbers to describe rotations. In fact, the Schrödinger wave equation – the foundational equation of quantum mechanics – contains the imaginary number i. The equation describes the probability of a subatomic particle being detected at a certain location. The probability of something happening must be between 0 and 1, or 0 and 100 per cent, of course. But the best way to understand how the probabilities of particles interact is to treat the probabilities as numbers on the complex plane. Rather than adding together like real numbers, the probabilities reinforce or cancel each other out depending on their relative positions in a rotation.

Thanks to equations like Schrödinger's, physicists now use imaginary numbers to describe the nature of matter itself. As a result, mathematicians no longer agonize about whether the imaginaries have external meaning or not. It is now as natural to think of, say, $2 + 3i$ existing on the complex plane as it is, say, -2 existing on the number line.

The complex plane brings us new insights into Euler's identity. To understand what they are, I need to introduce an alternative coordinate system for complex numbers. The standard system, as we have seen, is to let the point (a, b), where a is distance along from 0 and b is distance up from 0, refer to the complex number $a + bi$. Our second system, which uses 'polar' coordinates, is to describe the point (a, b) as being at angle θ and distance r from the origin. This is just like the commander of a submarine in an action movie who announces that there is a ship r miles away at a bearing of θ degrees (except that we are measuring our angle in radians, and measuring anticlockwise from east, rather than clockwise from north). In the illustration below, the dot represents the complex number $a + bi$. I have marked the angle θ from the horizontal and the distance r from the origin, which produces a right-angled triangle with angle θ, hypotenuse r, adjacent side a, and opposite side b.

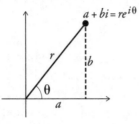

Soh-cah-toa!

The trigonometer's clarion call reminds us that sine = $\frac{\text{opposite}}{\text{hypotenuse}}$, and cosine = $\frac{\text{adjacent}}{\text{hypotenuse}}$. Which in this case means that:

$$\sin \theta = \frac{b}{r}, \quad \text{and} \quad \cos \theta = \frac{a}{r}$$

Which we can rearrange as:

$$b = r \sin \theta, \quad \text{and} \quad a = r \cos \theta$$

Our complex number can therefore be rewritten in terms of r and θ:

$$a + bi = r \cos \theta + (r \sin \theta)\, i$$
$$a + bi = r \cos \theta + ri \sin \theta$$
$$a + bi = r\, (\cos \theta + i \sin \theta)$$

Wait! We know that $\cos \theta + i \sin \theta = e^{i\theta}$, from the equation on p. 181, which means we can substitute the terms in brackets to get:

$$a + bi = re^{i\theta}$$

Savour this equation for a moment. The complex number that is r from the origin at an angle of θ radians from the horizontal has the form $re^{i\theta}$. Earlier in this chapter I asked what it meant for e to have an imaginary power, since it appeared baffling. Our answer is here. When e has an imaginary power, the term provides a fantastically efficient, submarine-commander-style notation for a position on the complex plane.

Now let's consider the coordinate (−1, 0) on the complex plane, which represents the complex number −1 + 0i, or simply −1. The point is 1 unit from the origin at an angle of π radians, as illustrated below, and therefore we can write it as $e^{i\pi}$.

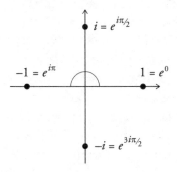

We have rediscovered Euler's identity! The statement describing the position of −1 on the complex plane is:

$$-1 = e^{i\pi}$$

Which rearranges to:

$$e^{i\pi} + 1 = 0$$

Furthermore, since the point i is 1 unit from the origin at an angle of $\frac{\pi}{2}$ radians from the horizontal, we can deduce that $i = e^{i\pi/2}$, and since the point $-i$ is 1 unit from the origin at an angle of $\frac{3\pi}{2}$ radians, we can deduce that $-i = e^{3i\pi/2}$.

Take a deep breath. We will now use this information to answer an eye-popping question that only a few pages ago would have seemed like insanity wrapped in madness: what is i^i, or (the square root of minus one) to the power of (the square root of minus one)?

Since we know that $e^{i\pi/2} = i$, we also know that:

$$i^i = (e^{i\pi/2})^i = e^{i^2\pi/2} = e^{-\pi/2} = \frac{1}{e^{\pi/2}} = \frac{1}{\sqrt{e^\pi}} = 0.20787\ldots$$

The i's pop out, leaving a term that even the Greeks would understand. Imagine that.

———

The complex plane allows us to forget the troubling notion that i is the square root of a negative number. All we need to be concerned with is that the complex number $a + bi$ represents the coordinates (a, b), where a and b are real numbers, and that when we add and multiply coordinates together they obey certain rules. (Of course, these rules rely on the properties of the square root of minus one, but the emphasis now is not on how they emerged but on what they do). Mathematicians soon began to ask whether it was possible to create rules for three-dimensional coordinates that would provide a way to describe rotations in space, in the same way that the rules for complex numbers described rotations in two. No one thought harder about this idea than the Irish mathematician William Rowan Hamilton, yet he was unable to find an answer. Then one day in 1843, as he was walking with his wife along the Royal Canal in Dublin, the solution came to him, and he performed the most famous act of vandalism in mathematics – he scraped the following formula on the wall of Brougham Bridge: $i^2 = j^2 = k^2 = ijk = -1$. A plaque now commemorates the spot.

Hamilton realized that it was impossible to find mathematically valid rules for coordinates with three numbers, but that he could make it work with four. He called his invention 'quaternions'. Just as the complex number $a + bi$, where a and b are real numbers and i is $\sqrt{-1}$, can be written as the coordinates (a, b), the quaternion $a + bi + cj + dk$, where a, b, c and d are real numbers and where i, j and k are all $\sqrt{-1}$, can be written (a, b, c, d). Each of the imaginary units i, j and k is equal to $\sqrt{-1}$, but nevertheless they are all different and are related by the equation in his graffiti. Hamilton required another bizarre rule for his quaternions to work – the order in which you multiply the imaginary units is important. For example, $i \times j = k$ but $j \times i = -k$.

Unorthodox though Hamilton's quaternions were, they nevertheless enabled him to produce a model for rotations in three dimensions. In the quaternion (a, b, c, d) the numbers (b, c, d) provide coordinate positions for the three spatial dimensions, and the number a, he said, was time. The new numbers so excited Hamilton that he devoted much of the rest of his life to studying them.

If you think that quaternions sound a bit strange, you're not the only one. Hamilton's peers ridiculed him, most notably Charles Dodgson, the Oxford maths don also known as Lewis Carroll. His children's books *Alice's Adventures in Wonderland* and *Through The Looking-Glass* are well known for their logical puzzles and mathematical games. Recently, however, one critic has argued that their surreal humour came not from Dodgson's florid imagination but from a desire to satirise changes in Victorian mathematics of which he disapproved, especially the trend for increasing abstraction in algebra. Melanie Bayley writes that the chapter 'A Mad Tea Party' lampoons Hamilton's quaternions, its title a pun on 'mad *t*-party', where *t* is the scientific abbreviation for time. At the party, the Hatter, the March Hare and the Dormouse are rotating around the table just like the imaginary numbers i, j and k in a quaternion. The fourth guest, Time, is absent, so there is no time for washing up. When the March Hare tells Alice that she should say what she means, she replies that 'at least I mean what I say – that's the same thing, you know'. Yet word order does change the meaning, just as the order of i and j in quaternion multiplication changes the result.

Dodgson sneered at the new mathematics, but he was on the wrong side of history. Hamilton's expansion of the concept of number to include quaternions broke the umbilical cord between numbers and meaning that had always previously existed. It is now uncontroversial for mathematicians to create new types of number based purely on formal definitions. A 'meaning' may be found – as complex numbers found meaning as positions in the complex plane – or it may not be. The purpose is to investigate pattern and structure and to see where you end up.

By the end of the nineteenth century, quaternions had been superseded by other mathematical theories, but Hamilton would be as happy as a leprechaun to know that for the last few decades they have been back in vogue. They provide computers with the best system to calculate roll, pitch and yaw, the three axes of rotation of an object in flight. Companies and organizations in the aeronautics and computer graphics industries – from Nasa to Pixar – all use them in their software.

It is impossible to create a consistent number system with five, six or seven ordered real numbers, but eight is fine, and this type of number is called an 'octonion', written (a, b, c, d, e, f, g, h). The octonion is an idea waiting for an application, although maybe not for long. One of the main contenders for a 'theory of everything', which reconciles quantum mechanics and general relativity, is M-theory, a version of string theory, in which the fundamental particles in an atom are taken to be strings. M-theory requires eleven dimensions, which some have argued are the eight dimensions of the octonions, plus the three dimensions of space. Hamilton scrawled his ideas on an Irish bridge, but they may have already been carved into the fabric of the cosmos.

Bertrand Russell, the only mathematician to win a Nobel Prize for literature, described beauty in maths as 'cold and austere, like that of sculpture, without appeal to any part of our weaker nature, without the gorgeous trappings of painting or music, yet sublimely pure, and capable of a stern perfection such as only the greatest art can show'. Euler's identity – pure, perfect and profound – fits the bill. Mathematical beauty can also be aesthetic, although Russell did not live to see one important example of how. In 1980, a decade after he died, a shape was discovered on the complex plane that was so striking and unusual it changed the way we think, not only about mathematics, but also about science.

Before we get there I need to introduce the concept of 'iteration', which is the process of repeating an operation again and again. We touched on this in the last chapter with the doubling sequence:

1, 2, 4, 8, 16, 32, 64, 128 ...

Instead of writing out the terms I could have defined the sequence as the iteration '$x \rightarrow 2x$', where the first term is 1, since:

$1 \rightarrow 2$
$2 \rightarrow 4$
$4 \rightarrow 8$
And so on.

What makes this process an iterative one is that the output of each operation – doubling, in this case – is used as the input for the subsequent operation. An iteration is a feedback machine: the number that comes out is fed back in, producing a new number, which is then fed back into the machine, and so on.

Now consider the simple iteration $x \rightarrow x^2$.

If we start at 1, we generate:

$1 \rightarrow 1^2 = 1$

$1 \rightarrow 1$

$1 \rightarrow 1$

In other words, the sequence stays at 1 for ever.

If we start at 2, we generate:

$2 \rightarrow 2^2 = 4$

$4 \rightarrow 16$

$16 \rightarrow 256$

$256 \rightarrow 65536 \rightarrow \ldots$

The sequence heads off to infinity.

And if we start at 0.1, we generate:

$0.1 \rightarrow (0.1)^2 = 0.01$

$0.01 \rightarrow 0.0001$

$0.0001 \rightarrow 0.00000001 \rightarrow \ldots$

The sequence bears down on zero.

We can generalize the behaviour of all numbers under this iteration. If the positive number n is larger than 1, its square n^2 is larger than n, so the number under iteration gets bigger and bigger. If a positive number n is smaller than 1, then n^2 is a fraction of n, so the number under iteration gets smaller and smaller and approaches zero. Since the square of a negative number is a positive number, all numbers less than −1 head to infinity, and all negative numbers between −1 and 0 head to zero.

I'm going to give the name 'escapee' to a number that zooms off to infinity, and the name 'prisoner' to one that doesn't. In the case of $x \rightarrow x^2$, we saw that 2 is an escapee, but that 1 and 0.1 are prisoners.

191

Our aim for the rest of this chapter will be to find the prisoners of any iteration, which we will call the 'prisoner set'. The prisoner set of $x \rightarrow x^2$ is the numbers between -1 and 1, as marked in bold in the illustration below.

The prisoner set of $x \rightarrow x^2$.

Consider a new iteration: $x \rightarrow x^2 + c$, where c is the starting number in the iteration. In other words, our feedback machine is a little bit hungrier than usual. It starts with the number c, squares it and adds c, squares the answer and adds c, squares the answer and adds c, and so on. This small modification to the rule has drastic consequences for determining which starting numbers are prisoners and which are escapees.

Let's start with 1, which as we saw above is a prisoner under the iteration $x \rightarrow x^2$. Under the iteration $x \rightarrow x^2 + c$, however, 1 is now an escapee:

Note: we start with 1, so $c = 1$
$1 \rightarrow 1^2 + 1 = 2$
$2 \rightarrow 2^2 + 1 = 5$
$5 \rightarrow 26$
$26 \rightarrow 677 \rightarrow 458330 \rightarrow \ldots$

Now let's consider what happens to -2, which is an escapee under the iteration $x \rightarrow x^2$. Under the iteration $x \rightarrow x^2 + c$, however, the number -2 is now a prisoner:

Note: we start with -2, so $c = -2$
$-2 \rightarrow -2^2 - 2 = 2$
$2 \rightarrow 2^2 - 2 = 2$
$2 \rightarrow 2$
$2 \rightarrow 2$
$\ldots$

It turns out that the prisoner set of $x \rightarrow x^2 + c$ contains the numbers between -2 and 0.25, as shown below in bold.

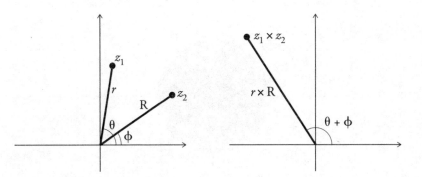

The prisoner set of $x \rightarrow x^2 + c$.

Let's now play prisoners vs escapees on the complex plane, the coordinate map where every point is defined by a complex number. Firstly, a brief recap about multiplication on the complex plane: multiplication by i is equivalent to an anticlockwise rotation by 90 degrees. More generally, when two complex numbers are multiplied by each other, the angles they make to the horizontal are added, and their distances from the origin are multiplied. (By convention we call a complex number z, rather than $a + bi$.) In the illustration below, the complex number z_1 is θ degrees from the horizontal at a distance of r, and z_2 is ϕ degrees from the horizontal at a distance of R. The complex number $z_1 \times z_2$ is therefore at an angle of $\theta + \phi$ degrees from the horizontal at a distance of $r \times$ R. We can now understand why multiplication by i is a quarter-turn. The position of i on the complex plane is $(0, 1)$, one unit up the imaginary axis, at a right angle from the horizontal. So, multiplication of a point on the complex plane by i rotates the point 90 degrees anticlockwise and multiplies the point's distance from the origin by 1, which means that the distance stays the same – the mathematical description of a quarter-turn.

Multiplication on the complex plane.

What happens to the complex numbers under the iteration $z \to z^2$?

Let's start with the imaginary unit i:

$i \to i^2 = -1$

$-1 \to 1$

$1 \to 1$

So i is in the prisoner set.

There is a quicker way to bring the prisoner set to light on the complex plane, using what we just learnt about complex multiplication. When we multiply two complex numbers, we add the angles and multiply the distances. So, when we square a complex number, we double its angle and square the distance. Consider the unit circle, which is the circle with radius 1 centred on the origin. All points on the unit circle are at a distance 1 from the origin, which means that the squares of any of these points are at a distance of $1^2 = 1$ from the origin. In other words, the square of a point on the unit circle remains on the unit circle, and so for the iteration $z \to z^2$ all of the points on the circle must be in the prisoner set. Similarly, if the distance from a point to the origin is less than 1, the square of that point is closer to the origin, and it will get closer and closer under iteration, so everything inside the unit circle is also in the prisoner set. But if the distance from the point to the origin is more than 1, the square of that point is further from the origin, and it will escape further and further under iteration. The prisoner set for the iteration $z \to z^2$ is the unit disc, illustrated below.

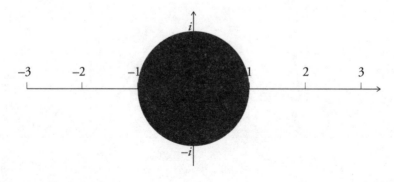

The prisoner set of $z \to z^2$.

Now hold on to your hats. We want to know the prisoner set of $z \rightarrow z^2 + c$, where c is the starting point in the iteration. Let's think about what the iteration means on the complex plane. We take a point, c, and then square it, which rotates it around the origin, and squares its distance from the origin. We then *add* c, which moves the point across the plane by c. This new point is then rotated, and its distance squared, before again it is moved across by c. The iteration is an endless alternating series of rotation, extension and translation; the plane is being turned, stretched and shifted at every point. There is no clever way of deducing what the prisoner set will look like. The only way is by iterating thousands and thousands of points, which was unfeasible before the computer age.

In 1979 Benoit B. Mandelbrot, a French mathematician working for IBM, became interested in $z \rightarrow z^2 + c$. His early printouts showed a blob-like prisoner set, but there were also tiny splotches which looked like specks of dust disconnected from the main blob. He left messages for his assistants saying that the imperfections were not a fault of the machine, to prevent them deleting the specks from the printouts. When Mandelbrot zoomed in on these areas he saw that they consisted of strikingly ornate patterns connected to the prisoner set by tiny branches. Gradually the overall picture of the prisoner set emerged. It looked like a sharp-snouted beetle with prickly fur, and was unlike any geometrical object ever seen before.

On a large scale the Mandelbrot set, as the shape is known, is ugly, even scary. But when you look closer, vistas of intricate beauty come into view. The series of eight images overleaf shows a zoom

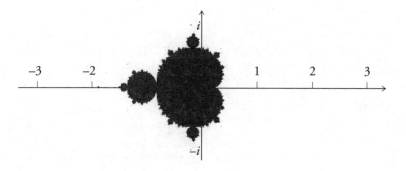

The prisoner set of $z \rightarrow z^2 + c$: the Mandelbrot set.

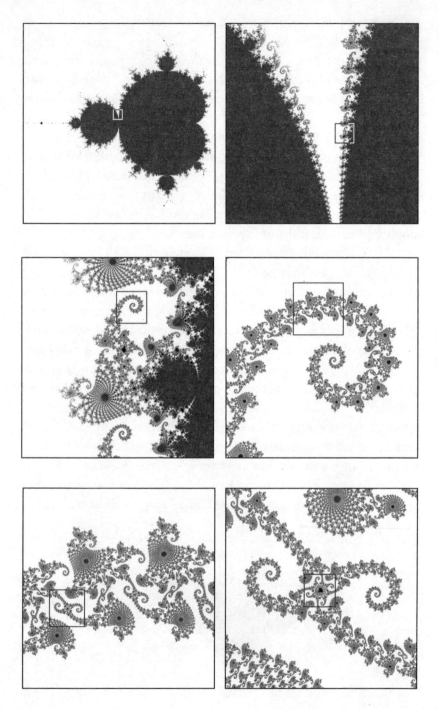

A journey into Seahorse Valley.

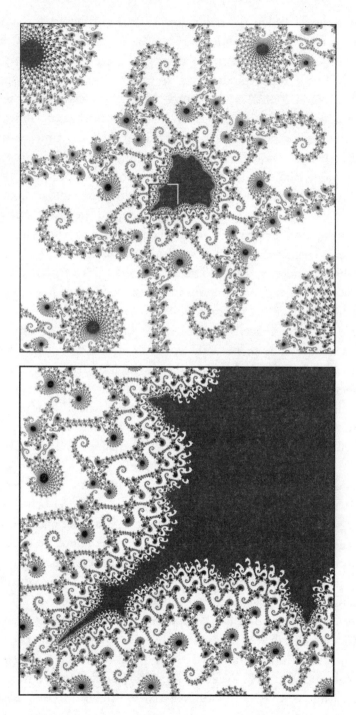

into Seahorse Valley, the name given to the area between the head and the body of the Mandelbrot set. The warty corrugations along the perimeter turn out to be paisley filigrees with seahorse-like spirals. Inside these spirals are more spirals, and then more spirals within spirals, until a miniature Mandelbrot-like set appears, embedded like an insect preserved in amber. 'It leaves us no way to become bored, because new things appear all the time, and no way to become lost, because familiar things come back time and time again,' Mandelbrot wrote. The variation is both infinitely deep and infinitely wide: wherever you look on the boundary a zoom will reveal an endlessly changing landscape. The battle between prisoners and escapees is so perfectly balanced that there are swirling skirmishes at every point and every scale.

The Mandelbrot set is a 'fractal', a word coined by Mandelbrot to mean any shape that contains miniature versions of itself. (A joke did the rounds that the B in Benoit B. Mandelbrot stood for Benoit B. Mandelbrot.) Fractals appear frequently in nature – a cauliflower floret has the same shape as the entire head, and a branch of a fern looks like the whole plant – and it is the Mandelbrot set's fractal properties that make its endlessly unfolding patterns look so organic. Mandelbrot's discovery was a rare moment when an advance in pure mathematics was equally an event in popular culture. The fractal was splashed across the covers of magazines and became a staple on bedroom walls, a bona fide 1980s icon like Adam Ant or shoulder pads. It still has its devotees. As computers have become more powerful, explorers have looked deeper than ever before, each voyage now an artistic and spiritual quest as much as a scientific one.

The Mandelbrot set can be made to look more beautiful by grading the escapees with different colours depending on how fast they head to infinity, and zooms can be animated to give the effect of falling through the world. Orson Wang, an automotive engineer from Detroit, bought three computers to zoom in further than any Mandelbrother had gone before. He spent three months choosing the best location to start, eventually going with a point near the complex number $-1.7 + 0.2i$, on the mini-Mandelbrot on the snout. He set his computers to work for six months to produce a zoom that magnifies 10^{275} times, more or less equivalent to zooming from the size of the observable universe into a proton six times. The result is mesmerizing. A sharp thorny spike transforms into a horizontal filament, a cross, an eight-pointed star, an irregular intersection of gnarled stems, and all of a sudden concentric circles explode in the middle like a swirling eye. It is impossible not to be enraptured. 'To me the Mandelbrot [set] epitomizes both insurmountable complexity and hope,' Orson said.

The discovery of the Mandelbrot set was a victory for computers, since it showed that they could help create new maths. It also marked a broader change in science. Before Mandelbrot, the accepted view was that the closer you looked at something, the simpler it became. Scientific enquiry aimed to break things down to their basic elements. Yet here was a shape that became more and more complex when looked at in finer and finer detail. Moreover, it showed that you could produce limitless complexity from a simple rule. The most amazing property of the Mandelbrot set is that it is created by multiplication and addition only, two basic operations understandable to a seven-year-old child.

The bottom of Seahorse Valley is the point −0.75. In 1991 the mathematician Dave Boll was trying to show that there are no prisoners directly above this point, so he started iterating points closer and closer to it. He started with $-0.75 + 0.1i$, and worked out how many iterations it took until the complex number in the sequence was more than two units from the origin, since once you get to two units from the origin, you are guaranteed that the original point is an escapee. He then did the same for $-0.75 + 0.01i$, and so on, to produce this table:

Imaginary part	Number of iterations	Imaginary part × iterations
0.1	33	3.3
0.01	315	3.15
0.001	3143	3.143
0.0001	31417	3.1417
0.00001	314160	3.14160

Can you see where the final value is heading? Closer and closer to pi.

Deep questions remain unanswered about the Mandelbrot set, but for many years the challenge that most interested recreational mathematicians was whether it was possible to create one in three dimensions. Since there is no three-dimensional model of the complex plane, there was no obvious solution. In 2009, however, Daniel White cracked the problem. A 31-year-old piano teacher from Bedford, he transferred the principles behind complex multiplication from two to three dimensions.

A point on a plane can be defined by its distance from the origin and its angle from the horizontal.

Likewise, a point in space can be defined by its distance from the origin and two angles, one from the horizontal axis and one from the vertical axis, in the same way that a point on the globe can be specified by a latitude (vertical angle) and a longitude (horizontal angle).

When you multiply two complex numbers, you add their angles and multiply their distances. Dan defined the multiplication of two points in space to be the addition of the two horizontal angles, the addition of the two vertical angles, and the multiplication of the distances.

With this definition, he looked at the 3-D prisoner set of $z \rightarrow z^2 + c$. The result was disappointing. 'It looked like whipped cream,' he said. The Mandelbrot set's infinite liveliness was subdued, and he commiserated with pals on an internet forum for fractal fans. The breakthrough came when the Los Angeles-based mechanical engineer Paul Nylander suggested that Dan use his method with the iteration $z \rightarrow z^8 + c$. The tweak transformed the whipped cream into a barnacle-encrusted fractal planet with deep caves, swirling mountains and star-shaped crevasses, now called the Mandelbulb. 'I was awestruck,' said Dan. 'It was like a new universe had been discovered.'

Awe is the only proper reaction to the Mandelbulb. The object could be a spaceship, a sea creature, an alien virus ... whatever you want it to be. Its surface is more detailed and fantastic than anything conceivable from a human imagination, yet the entire structure is precisely determined by a single line of code: $z \rightarrow z^8 + c$.

The discovery of the Mandelbrot set reconnected mathematics with the natural sciences. Fractal geometry was a new approach to understanding complex forms in natural phenomena, from weather systems to coastlines, and from organisms to crystals. Three hundred years previously, another mathematical breakthrough had an even greater impact on how we see the world around us.

Professor Calculus

The French mathematician Cédric Villani is no ordinary-looking university professor. Handsome and slender, with a boyish face and a wavy, neck-length bob, he looks more like a dandy from the belle époque, or a member of a pretentious student rock band. He always wears a three-piece suit, starched white collar, *lavaliere* cravat – the kind folded extravagantly in a giant bow – and a sparkling, tarantula-sized spider brooch. 'Somehow I had to do it,' he said of his appearance. 'It was instinctive.'

I first met Villani in Hyderabad, India, at the 2010 International Congress of Mathematicians, or ICM, the four-yearly gathering of the tribe. Of the three thousand delegates, Villani was the focus of most attention, not because he was the most elaborately dressed, but because he received the Fields Medal at the opening gala. The Fields is the highest honour in maths. Villani was living the life of a pop star, unable to cross a room without being accosted for autographs and pictures. I grabbed him one afternoon and asked if celebrity mathematicians attract groupies. 'You know, in mathematics they are a bit shy, so they will not assault me,' he laughed. 'Unfortunately.'

Fields Medals are awarded at each ICM to two, three or four mathematicians under the age of 40. (At Hyderabad, the other medallists were Elon Lindenstrauss from Israel, Stanislav Smirnov from Russia and Ngô Bảo Châu from Vietnam.) The age rule recognizes the original motivation behind the prize, which was conceived by the Canadian mathematician J. C. Fields. He wanted not just to recognize work already done, but also to encourage future success. Such is the acclaim afforded by a Fields Medal, however, that since the first two were awarded in 1936, they have helped establish a cult of youth, implying that once you hit 40 you're past it. This is unfair. Many mathematicians produce their best work after the age of 40, although Fields medallists can struggle to regain focus, since fame brings with it other responsibilities. The maths community does not honour lifetime achievement in the way that the physics and chemistry communities do with the Nobel Prize.

The first ICM was held in 1897 in Zurich. At the second, in 1900 in Paris, the German David Hilbert gave a lecture in which he listed 23 unsolved mathematical problems, which determined the direction of the discipline for the next hundred years. Mathematicians gather at the ICM to take stock of their achievements, and the Fields Medal citations provide the clearest snapshot of the most exciting recent work. Lindenstrauss was commended 'for his results on measure rigidity in ergodic theory, and their applications to number theory'; Smirnov 'for the proof of conformal invariance of percolation and the planar Ising model in statistical physics'; and Ngô 'for his proof of the Fundamental Lemma in the theory of automorphic forms through the introduction of new algebro-geometric methods'. You may be as bewildered as I was when the citations were announced at the congress – in fact, many of the delegates were too, even during the subsequent explanatory lectures. The British mathematician Timothy Gowers, a Fields laureate from 1998, wrote on his blog: 'If you want to impress your friends, here's how to pretend you understand [Ngô's work]. If someone asks what his main idea was, you can reply, "Well, his deepest insight was to show that the Hitchin fibration of the anisotropic part of the trace formula is a Deligne-Mumford stack." If that doesn't do the job, then try to drop the phrase "perverse sheaves" into the conversation – they are relevant apparently.' At its cutting edge, mathematics is so conceptually difficult that maybe only a few hundred people in the world can understand what each Fields medallist did. With Ngô, the most abstract thinker, perhaps even fewer.

Villani's citation, however, was not as impenetrable as the others. He won 'for his proofs of nonlinear Landau damping and convergence to equilibrium for the Boltzmann equation'. Here, at least, was something understandable to the non-specialist. The Boltzmann equation, devised by the Austrian physicist Ludwig Boltzmann in 1872, concerns the behaviour of particles in a gas, and is one of the most famous equations in classical physics. Not only is Villani a devotee of the nineteenth century's neckwear, he is also a world authority on its applied mathematics.

The Boltzmann equation is what is known as a partial differential equation, or PDE, and it looks like this:

$$\frac{\partial f}{\partial t} + v \cdot \nabla_x f = \int_{\mathbb{R}^3} \int_{\mathbb{S}^2} |v - v_*| \left[f(v')f(v'_*) - f(v)f(v_*) \right] dv_* d\sigma$$

If you studied calculus at school you may recognize some of the symbols, especially the extended $\int$ or the curly ∂. If you didn't study calculus, don't be afraid, as I am about to explain what they mean. Calculus was the crowning intellectual achievement of the Enlightenment, and Villani's Fields Medal demonstrates that it remains a rich area of advanced mathematical study. We will return to the flamboyantly attired Frenchman and his equation, but to equip ourselves with the requisite conceptual and terminological tools, we first need to transport ourselves from southern India in 2010 to Sicily in around the third century BCE.

On the front of the Fields Medal is the bearded portrait of Archimedes, basking in the glow of his reputation as the most illustrious mathematician of antiquity. Archimedes, however, is usually remembered for his contributions to physical science. When he leapt out of his bath shouting 'Eureka!', for example, he was not celebrating a mathematical discovery, but a breakthrough in fluid mechanics: the principle of buoyancy. His best-known inventions included a giant claw that destroyed ships invading Syracuse, his hometown, and a screw that raised water when turned by hand. Yet the historian Plutarch wrote that Archimedes considered engineering 'sordid and ignoble'. Geometry was the object of his 'whole affection and ambition'. At bath times not disturbed by thoughts of physics, 'he would ever be drawing out of the geometrical figures, even in the very embers of the chimney. And while [his servants] were anointing of him with oils and sweet savours, with his fingers he drew lines upon his naked body, so far was he taken from himself, and brought into ecstasy or trance, with the delight he had in the study of geometry.'

The initial task of geometry was the calculation of area. (The historian Herodotus was the first to use the word 'geometry', or earth-measure, describing it as a practice devised by Egyptian tax inspectors to calculate areas of land destroyed by the Nile's annual floods.) As we all know, the area of a rectangle is the width multiplied by the height, and from this formula we can deduce that the area of a triangle is half the base times the height. The Greeks devised methods to calculate the areas of more complicated shapes. Of these, the most impressive achievement was Archimedes's 'quadrature of the parabola', by which is meant the calculation of the area bounded by a parabola and a line. Archimedes first drew a large triangle inside the parabola, as illustrated below, then on either side of this he drew another triangle. On each of the two sides of these smaller triangles, he drew an even smaller triangle, and so on, such that all three points of each triangle were always on the parabola. The more triangles he drew, the closer and closer their combined area was to the area of the parabolic section. If the process was allowed to carry on for ever the infinite number of triangles would perfectly cover the desired area.

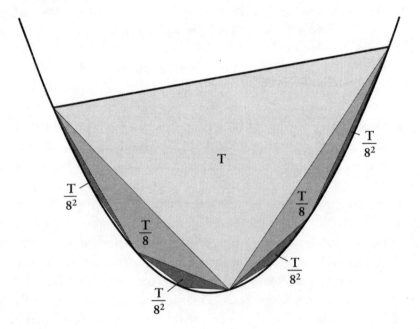

The quadrature of the parabola.

Archimedes continued by demonstrating that if the area of the large triangle is T, the area of each of the two triangles drawn on its sides is $\frac{T}{8}$, the area of each of the four triangles drawn on their sides is $\frac{T}{8^2}$, and so on. In other words, the area of the parabolic section, which is the combined sum of all the triangles, is the infinite series:

$$T + \frac{2T}{8} + \frac{4T}{8^2} + \frac{8T}{8^3} + ...$$
or
$$T(1 + \frac{1}{4} + \frac{1}{16} + \frac{1}{64} + ...)$$
or
$$T(1 + \frac{1}{4} + \frac{1}{4^2} + \frac{1}{4^3} + ...)$$

To finish off, he proved that this series was equal to $\frac{4T}{3}$. So, if we want to measure the area between a line and a parabola, we draw a triangle, measure the base and height, calculate the area, and multiply the answer by $\frac{4}{3}$. I won't give Archimedes's proof of this last part but instead show you a picture that contains a proof within it. This type of mathematical diagram is called a 'proof without words' and the illustration below is probably my favourite in this book. It says that:

$$\frac{1}{3} = \frac{1}{4} + \frac{1}{4^2} + \frac{1}{4^3} + ...$$

Look at it for a while to see if you can work out why. (And if you can't, see Appendix Six on p. 299.) If this equation is true, then the total area $T(1 + \frac{1}{4} + \frac{1}{4^2} + \frac{1}{4^3} + ...) = T(1 + \frac{1}{3}) = \frac{4T}{3}$. QED.

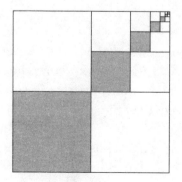

Archimedes's quadrature of the parabola is the most sophisticated example from the classical age of the *method of exhaustion,* the technique of adding up a sequence of small areas that converge towards a larger one. The proof is considered his finest moment because it represents the first 'modern' view of mathematical infinity. Two hundred years before Archimedes, the philosopher Zeno had warned against infinity in a series of paradoxes. The most famous of these, Achilles and the Tortoise, showed that adding an infinite number of quantities together leads to nonsense.

Imagine, said Zeno, that Achilles is racing to catch up with a tortoise. When the athlete reaches the place where the tortoise was when he started his run, the animal has moved a little bit forward. When he reaches this second position, again the tortoise has advanced. Achilles can carry on for as long as he likes, but whenever he reaches the place where the tortoise was at the beginning of his most recent dash, the animal is still ahead. By treating motion as an infinite number of sprints over an infinite number of intervals, Zeno argued that swift Achilles will never reach his sluggish rival. The Greeks were never able to untangle Zeno's logical knots, and as a result mathematicians avoided infinity in their work. Even Archimedes, when he used the method of exhaustion, never referred as crudely as I just did to a complete entity called an 'infinite series'. Still, the difference was in terminology rather than ideas. Archimedes was the earliest thinker to develop the apparatus of an infinite series with a finite limit. This was important not just for conquering the areas of shapes significantly more exotic than the parabola, but also for starting on the conceptual path towards calculus. Of the giants on whose shoulders Isaac Newton would eventually perch, Archimedes was the first.

If infinity is the biggest number, what is the smallest? In the seventeenth century mathematicians introduced a new concept called the 'infinitesimal', a quantity that was smaller than any real amount, yet still larger than zero.

The infinitesimal was both something and nothing: large enough to be of mathematical use, but small enough to disappear when you needed it to. For example, consider the circle opposite.

Inside is a dodecagon, a 12-sided shape made up of 12 identical triangles sharing a common vertex, or point. The combined area of the triangles is approximately the area of the circle. If I drew a polygon with more sides within the circle, containing more, thinner triangles, their combined area would approximate the circle more closely. And if I kept on increasing the number of sides, in the limit I would have a polygon with an infinite number of sides containing an infinite number of infinitely thin triangles. The area of each triangle is infinitesimal, yet their combined area is the area of the circle.

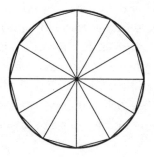

We have met the German astronomer Johannes Kepler twice already. He was the man who realized that planets orbit in ellipses, and he was also the man who went on eleven dates before finding a second wife. Once he had proposed to the future Frau K., there was the small matter of organizing the wedding. While purchasing the booze, he saw that the vintners evaluated how big a full barrel of wine was by inserting a rod diagonally through the bunghole halfway along, and then measuring it when it hit the far edge. It was a crude, approximate method, distasteful to Kepler, since the same length corresponded with many different sizes of barrel, as shown overleaf.

He started thinking about how to calculate volumes more accurately, to determine the shape that carried the most wine for a fixed length of rod. Inspired by Archimedes, he developed a method in which he divided each barrel into an infinite number of infinitesimally small volumes. He then proved that for a rod of length l going from bunghole to far corner, the barrel has maximum volume if its width is $\frac{2l}{\sqrt{3}}$. Kepler was the trailblazer

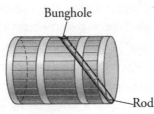
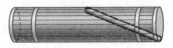

Fat barrel   Thin barrel

Measuring the size of wine barrels.

for a whole generation of mathematicians who used infinitesimals when calculating areas and volumes. From England to Italy there was an explosion of activity in this field, and it reflected the most significant change in mathematical culture since the ancient Greeks, who had been sticklers for concepts that made logical sense. Now, logical rigour was abandoned in favour of what got results. The infinitesimal was neither fish nor fowl, something that both did and didn't exist. But no one was going to give it up. It was just too powerful.

Infinitesimals provided an extremely successful method for calculating tangents, which are those lines that touch a curve at a single point, kissing rather than cutting it. Imagine we want to find the tangent at point P on the curve illustrated below. The strategy is to make an approximation of the tangent, and then to improve the approximation until it coincides with the desired line. We do this by drawing a line through P that cuts the curve at a nearby point Q, and then we bring Q closer and closer to P. When Q hits P, the line is the tangent.

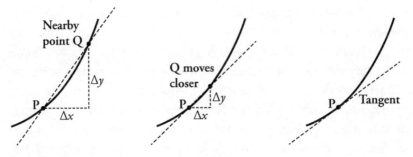

Approximating a tangent.

210

As we saw earlier, the gradient of a straight line is the distance it moves up divided by the distance it moves along, and the gradient of a point on a curve is the gradient of the tangent at that point. Mathematicians were interested in tangents because they were interested in gradients. In the illustration, the gradient of the line between P and Q is $\frac{\Delta y}{\Delta x}$. (The Greek letter delta, Δ, is a mathematical symbol meaning a small increment.) As Q closes in on P, the value of $\frac{\Delta y}{\Delta x}$ approaches the gradient of the tangent at P. But we have a problem. If we let Q actually reach P, then $\Delta y = 0$ and $\Delta x = 0$, meaning that the gradient of the curve at P is $\frac{0}{0}$. Bad maths alert! The rules of arithmetic prohibit division by zero! The solution is to keep Q at an infinitesimal distance from P. If we do, we can say that when Q becomes *infinitesimally* close to P, the value of $\frac{\Delta y}{\Delta x}$ is *infinitesimally* close to the gradient of the curve at P.

In 1665, Isaac Newton, recently graduated from Cambridge, returned to live with his mother in their Lincolnshire farmhouse. The Great Plague was devastating towns across Britain. The university had closed down to protect its staff and students. Newton made himself a small study and started to fill a giant jotter he called the Waste Book with mathematical thoughts. Over the next two years the solitary scribbler, undistracted, devised new theorems that became the foundations of the *Philosophiæ Naturalis Principia Mathematica*, his 1687 treatise that, more than any work before or since, transformed our understanding of the physical universe. The *Principia* established a system of natural laws that explained why objects, from apples falling off trees to planets orbiting the Sun, move as they do. Yet Newton's breakthrough in physics required an equally fundamental breakthrough in maths. He formalized the previous half-century's work on infinitesimals into a general system with a unified notation. He called it the *method of fluxions,* but it became better known as the 'calculus of infinitesimals', and now, simply, 'calculus'.

A body that moves changes its position, and its *speed* is the change in position over time. If a body is travelling with a fixed speed, it changes its position by a fixed amount every fixed period. A car with constant speed that covers 60 miles between 4pm and 5pm is travelling at 60 miles per hour. Newton wanted to solve a different problem: how does one calculate the speed of a body that is not travelling at a constant speed? For example, let's say the car above, rather than travelling consistently at 60mph, is continually slowing down and speeding up because of traffic. One strategy to calculate its speed at, say, 4.30pm, is to consider how far it travels between 4.30pm and 4.31pm, which will give us a distance per minute. (We just need to multiply the distance by 60 to get the value in mph.) But this figure is just the average speed for that minute, not the instantaneous speed at 4.30pm. We could aim for a shorter interval – say, the distance travelled between 4.30pm and 1 second later, which would give us a distance per second. (We'd then multiply by 3600 to get the value in mph.) But again this value is the average for that second. We could aim for smaller and smaller intervals, but we are never going to get the instantaneous speed until the interval is tinier than any other – when it is zero, in other words. But when the interval is zero, the car does not move at all!

This line of reasoning should sound familiar, because I used it three paragraphs ago when explaining how to calculate the gradient of a point on a curve. To find the gradient we divide an infinitesimally small quantity (length) by another infinitesimally small quantity (another length). To get the instantaneous speed we also divide an infinitesimally small quantity (distance) by another infinitesimally small quantity (time). The problems are mathematically equivalent. Newton's method of fluxions was a method to calculate gradients, which enabled him to calculate instantaneous speeds.

Let's see how Newton used this method to calculate the gradient of the curve $y = x^2$, our old friend the parabola. It's a bit technical, but not difficult if you go slowly. By the end you'll see how he used infinitesimals to find a formula for the gradient of every point along the curve.

We will proceed as we did in the illustration on p. 210, taking an arbitrary point P, and then approximating the tangent with a line through another point Q that is only a small distance along the curve. We'll then bring Q infinitesimally close to P. The gradient of the tangent at P is the gradient of the curve at P. To begin, we introduce a new quantity, o, which is the horizontal distance between

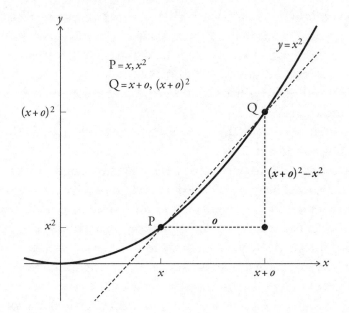

$P = x, x^2$

$Q = x + o, (x + o)^2$

$y = x^2$

$(x + o)^2$

$(x + o)^2 - x^2$

x^2

P

Q

o

x

$x + o$

Calculating the gradient of y = x².

P and Q, as shown above. If the coordinates of P are (x, x^2), then the coordinates of Q are $(x + o, (x + o)^2)$. The vertical distance between P and Q is therefore $(x + o)^2 - x^2$, so the gradient of the line – the vertical distance divided by the horizontal distance – is:

$$\frac{(x + o)^2 - x^2}{o}$$

Which expands to:

$$\frac{x^2 + 2xo + o^2 - x^2}{o}$$

Which cancels out to:

$$\frac{2xo + o^2}{o}$$

Which divides out to:

$$2x + o$$

As Q gets infinitesimally close to P, o becomes infinitesimally small, so the gradient becomes infinitesimally close to $2x$. Newton said that we are allowed to let Q coincide with P, and that when it does we can disregard the infinitesimal o, and state confidently that the gradient at P is $2x$. Once the infinitesimal has done its job, it can leave the scene.

In other words, for the curve $y = x^2$, the gradient at horizontal position x is equal to $2x$.

Even if you found the algebra here too complicated, you can still appreciate what Newton achieved. He extracted the most important property of the curve – its gradient – and derived a formula, $2x$, that lets us calculate it at any point along the curve. If we let y' be the symbol for gradient, we have produced a new equation: $y' = 2x$, which is also known as the 'derivative' of the original curve.

In the top left illustration opposite I have drawn the original curve $y = x^2$, and, directly below it, $y' = 2x$, its gradient, which is a straight line. When x is 1, the curve has value 1, and the gradient is 2. When x is 2, the curve has value 4 and the gradient is 4. The curve rises in the shape of a parabola, and the gradient rises linearly. Now forget about geometry and think about mechanics. The two images could equally describe the behaviour of a moving object. If the original curve plots the position of an object over time, then the derivative plots the object's instantaneous speed. The graphs show that when 1 unit of time has elapsed, the object has moved 1 unit, and the speed is 2. When 2 units of time have elapsed, the object has moved 4 units and the speed is 4, and so on. The top curve, in fact, models the position of an object when it falls freely under gravity: distance covered is proportional to the square of the time elapsed. Using calculus, Newton showed that the instantaneous speed of a falling object increases linearly as it drops.

I chose the curve $y = x^2$ because the derivative is straightforward to compute, but Newton's method can be applied to all smooth curves, provided we have an equation for them. Illustrated on the top right is another curve, together with the curve of its gradient, or derivative. I've left out the equation for these curves, however, and have simply called them A and B, because I want you to marvel at the poetry of the transformation. At every point along A, the

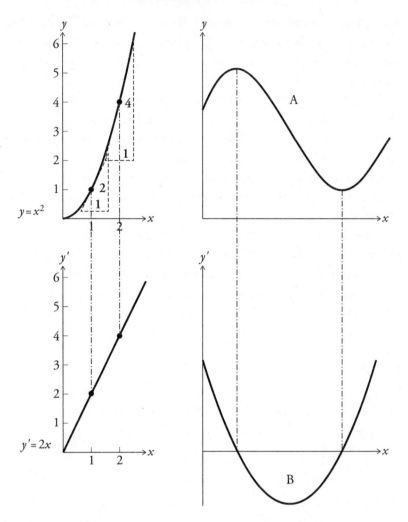

The gradient of the parabola on the top left is a straight line, and the gradient of curve A is curve B.

gradient is charted directly below on B. Let's travel along A from left to right. The curve rises, peaks, descends, bottoms out and then rises again. In other words, the gradient is positive, hits zero when the curve is momentarily horizontal, is negative, rises to zero and becomes positive again. This chronology is exactly what happens to B! It starts positive, crosses the horizontal axis into negative territory, and then bursts back onto the positive side. (The dashed vertical lines show how the landmarks of the top curve correspond

215

to the positions of zero gradient.) When I first saw a curve like this together with the curve of its gradient I was transfixed. That a quantity varying in one curve could be captured so perfectly by another curve seemed like magic.

The infinitesimal provided a method to find gradients. It also provided a method to find areas. We saw earlier how Archimedes calculated the area bounded by a parabola and a line by adding smaller and smaller triangles, and how Renaissance mathematicians refined this technique by dividing areas into subsections of infinitesimal size. Newton's method of fluxions provided a way to calculate the area underneath a curve, by dividing the area into an infinite number of infinitely thin vertical strips.

For example, if we know the equation of the curve C below, using calculus we can work out an equation for the shaded area A between the origin and x on the horizontal axis.

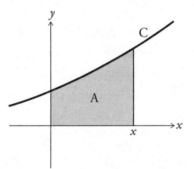

Given a curve, therefore, calculus gives us two options: we can find an equation for its gradient, or we can find an equation for the area underneath it. Yet here's the thing: these procedures are the reverse of each other! Gradient and area are the same thing seen from different directions. It's a plot twist worthy of Scooby-Doo – two separate characters in the drama of mathematics are in the final act revealed to be the same. The result is called the Fundamental Theorem of Calculus, and it was one of the most surprising discoveries of the seventeenth century.

Broadly speaking, the theorem states that if the area under curve

C is A, then the gradient of curve A is C. If this sounds confusing, remember that curves, areas and gradients are all written as equations. C is a curve and also an equation. Using calculus we can find the equation A for the area underneath it. The Fundamental Theorem of Calculus tells us that the derivative, or gradient, of the equation A is C.

Let's see how this works when C is the line $y = 2x$, illustrated below left. The formula for the area of a triangle is half the base times the height. (We could have worked this out with infinitesimals, but we don't need to as we know it already.) So, the area A underneath the line from 0 to x is $\frac{1}{2}x \times 2x$, or x^2, which gives us the equation $A = x^2$ for the area underneath the line. This equation is also the curve below right, the parabola. Recall the diagram on p. 215, which shows how by considering a curve's gradient, we go from the parabola to the straight line. The illustrations below show how by considering the area under a curve, we go from the straight line to the parabola. Gradient and area are flipsides of the same coin.

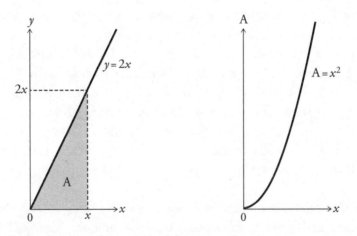

Calculating the area under y = 2x, and drawing it as a curve.

Calculus allowed Newton to take an equation that determined the position of an object, and from it devise a secondary equation about the object's instantaneous speed. It also allowed him to take an equation determining the object's instantaneous speed, and from it devise a secondary equation about position. Calculus gave him the mathematical tools to develop his laws of motion. In his equations, he called the variables x and y 'fluents' and the gradients 'fluxions', written as the 'pricked letters' $\dot{x}$ and $\dot{y}$.

When Newton returned to Cambridge after two years avoiding the plague in Lincolnshire, he did not tell anyone about the method of fluxions, a decision he came to rue. On the continent, Gottfried Leibniz was developing an equivalent system. Leibniz was German by birth but a man of the world – a lawyer, diplomat, alchemist, engineer and philosopher. Leibniz was also the mathematician most obsessed with notation. The symbols he used for his system of calculus were clearer than Newton's, and are the ones we use today.

Leibniz introduced the terms dx and dy for the infinitesimal differences in x and y. The gradient, which is one infinitesimal difference divided by the other, he wrote $\frac{dy}{dx}$. Thanks to his use of the word 'difference', the calculation of gradient became known as 'differentiation'. Leibniz also introduced the distinctive stretched 's', $\int$, as the symbol for the calculation of area. It's an abbreviation of *summa*, or sum, since as we have seen the calculation of area is based on infinite sums of infinitesimals. On the suggestion of his friend Johann Bernoulli, Leibniz called his technique *calculus integralis*, and the calculation of area became known as 'integration'. The benefit of such a long (and extendable) symbol is that it has space to include the two horizontal values that mark out the area being calculated. The area A, as illustrated in the figure on p. 216, is written $\int_0^x C$, and pronounced 'the integral of C from 0 to x'. Leibniz's $\int$ is the most majestic symbol in maths, reminiscent of the f-hole of a cello or violin.

For more than two decades Leibniz and Newton corresponded amicably and respectfully with each other about infinitesimals. When Leibniz published details of his calculus first, everyone assumed he had devised it independently. Then, in 1699, a few years after Newton went public with fluxions, a young Swiss mathematician living in England accused Leibniz of stealing Newton's ideas. It was five years before a response appeared: an article (possibly written by Leibniz) in the *Acta Eruditorum* suggesting that Newton had instead plagiarized Leibniz. The tit for tat between the British and the continental scientific communities turned increasingly nasty, and the feud dominated the latter years of both men's lives. Priority disputes were common in those days, but none involved men of such stature as Newton and Leibniz, nor were as bitter or enduring. When both men died, the rancour remained. Britain – where as a matter of national pride Newton's fluxions were used instead of differences – became isolated from European advances for the best part of a century. Only when the English adopted Leibniz's notation – passing from 'the dot-age of fluxions to the de-ism of the calculus', as Augustus de Morgan wrote – did Britain regain its status in mathematics.

In 1891 the German company Bahlsen began producing the rectangular, ribbed-edge butter biscuit Leibniz, named after Hanover's most famous son. Coincidentally, that same year, a baker in Philadelphia made the first Fig Newton, a pastry filled with fig paste and named after the town of Newton in Massachusetts. Nowadays, Newton vs Leibniz is an argument reserved for afternoon tea.

As we saw earlier, calculus consists of two procedures: differentiation (computation of gradient) and integration (computation of area). In general terms, gradient is the rate of change of one variable quantity over another, and area is the measure of how much one variable quantity accumulates with respect to another. Calculus thus provides scientists with a way to model the behaviour of quantities that change in relation to each other. It is a formidable instrument to explain the physical world because everything in the universe, from the tiniest atoms to the largest galaxies, is in a state of permanent flux.

When we know the relationship between two varying quantities, we can describe them in an equation using the symbols for differentiation and integration. An equation in x and y that includes the term $\frac{dy}{dx}$ is called a 'simple differential equation'. If there are more than two variables, say x, y and t, the rates of change are written $\frac{\partial y}{\partial x}$, or $\frac{\partial y}{\partial t}$, with the rounded ∂. The equation is called a 'partial differential equation', or PDE, because terms like $\frac{\partial y}{\partial x}$ tell us how one variable changes with respect to another one, but not to all of them. PDEs dominate applied mathematics. They allow scientists to make predictions. If we know how two quantities vary over time, then we can predict exactly what state they will be in at any time in the future. Maxwell's equations, which explain the behaviour of magnetic and electric fields, the Schrödinger equation, which underlies quantum mechanics, and Einstein's field equations, which are the basis of general relativity, are all PDEs.

The first important PDE described the behaviour of a violin string when bowed, a problem that had tormented scientists for decades. It was discovered in 1746 by Jean le Rond d'Alembert, the celebrity mathematician of his day. D'Alembert, the product of a brief liaison between an artillery general and a lapsed nun, was abandoned after he was born and left on the steps of the church Saint Jean Le Rond, next to Notre-Dame Cathedral in Paris, from which he took his name. Brought up by the wife of a glazier, he rose against the odds to become the permanent secretary of the Académie Française. As well as being a serious mathematician, he was also a vociferous apologist for the values of the Enlightenment. He was a public figure, a sought-after guest at aristocratic salons and one of the editors of the landmark *Encyclopédie*, for which he wrote the preliminary discourse and more than a thousand articles.

D'Alembert was the prototype French scientific intellectual, a role now occupied with gusto by Cédric Villani.

The second time I met Villani was in Paris. Since 2009 he has been director of the Institut Henri Poincaré, France's elite maths institute, which is situated among the universities of the Latin Quarter. His office is a comfortable clutter of books, paper, coffee mugs, awards, puzzles and geometrical shapes. Villani's appearance was unchanged since our first encounter at the International Congress of Mathematicians two years before: burgundy cravat, blue three-piece suit, and a metal spider glistening on his lapel. He said his look emerged when he was in his twenties. He wore shirts with large sleeves, then with lace, then a top hat ... 'It was like a scientific experiment, and gradually it was, "this is me".' And the spider? He enjoys its ambiguity. 'Some people think the spider is a maternal symbol. Others think that the web is a symbol for the universe, or that the spider is the big architect of the world, like a way to personify God. Spiders don't leave people indifferent. You immediately have a reaction.' The spider is an archetype rich with interpretations, I thought, just like mathematics is an abstract language with innumerable applications.

Villani's field is PDEs. Even though PDEs have been around for almost three centuries, he says they are 'for a large part still poorly understood. Each PDE seems to have a theory of its own. You have many sub-branches of PDEs with only a small common basis and no general classification. People have tried to classify them, but even the best specialists have failed.' The PDE that has occupied most of Villani's time is the Boltzmann equation. It was the subject of his PhD and formed part of the subsequent work that led to his Fields Medal. He now views it with tenderness and devotion. 'It's like the first girl you fall in love with,' he confided. 'The first equation you see – you think it is the most beautiful in the world.' Feast your eyes on her again:

$$\frac{\partial f}{\partial t} + v \cdot \nabla_x f = \int_{\mathbb{R}^3} \int_{\mathbb{S}^2} |v - v_*| \left[f(v') f(v'_*) - f(v) f(v_*) \right] dv_* d\sigma$$

The Boltzmann equation belongs to the field of statistical mechanics: the branch of mathematical physics that investigates how the microscopic behaviour of individual molecules influences macroscopic properties like temperature and pressure. The equation describes how a cloud of gas disseminates by considering the likelihood of any of its molecules being in any particular spot, with a particular speed, at a particular time. The model assumes that particles in a gas bounce around according to Newton's laws, but in random directions, and it describes the effects of their collisions using the maths of probability. Villani pointed at the left side of the equation: 'This is just particles going in straight lines.' He pointed to the right side of the equation: 'And this is just shock. Tik-ding! Ting-dik!' He bumped his fists together several times. 'Often in PDEs, you have tension between various terms. The Boltzmann equation is the perfect case study because the terms represent completely different phenomena and also live in completely different mathematical worlds.'

If you filmed a single gas particle bouncing off another gas particle, and showed it to a friend, there is no way he or she would know whether you were playing the film forwards or backwards, since Newton's laws are time-reversible. But if you filmed a gas spreading from a beaker to its surroundings, a viewer would instantly be able to tell which way the film was being played, since gases do not suck themselves back into beakers. Boltzmann established a mathematical foundation for the apparent contradiction between micro- and macroscopic behaviour by introducing a new concept, entropy. This is the measure of disorder – in theoretical terms the number of possible positions and speeds of the particles at any time. Boltzmann then showed that entropy always increases. Villani's breakthrough paper concerned just how fast entropy increases before reaching the totally disordered state.

The Boltzmann equation has straightforward applications, such as in aeronautical engineering, to determine what happens to planes when they fly through gases. Its usefulness is what first appealed to Villani when he embarked on his PhD. But as he became more intimate with the equation, its beauty seduced him. He compares it to a Michelangelo sculpture: 'Not pure and ethereal and elegant, but very human, very tortured, with the strength of the energy of the world. In the equation you can hear the roar of the particles, full of fury.' He added that he prefers to spend years studying well-known equations, trying to find new insights into them, rather than inventing new concepts. 'It's what I like, and it's part of a general attitude that says, "Hey, guys! High-energy physics, the Higgs boson, string theory or whatever – it may all be fascinating, but remember we still don't understand Newtonian mechanics." There are still many, many open problems.' He showed me a PDE in a book. 'Does this equation have smooth solutions? Nobody in hell knows that!' He shrugged his shoulders, his forehead criss-crossed with lines.

On the wall behind Villani is a poster of his favourite singer, sultry prog rock chanteuse Catherine Ribeiro, arms stretched apart, fists clenched. On his desk is a bust of the French mathematician Henri Poincaré, bearded and sombre. 'This is the duality of making things move, and thinking,' he quipped. Poincaré, who lived at the turn of the nineteenth century, had a reputation as the last mathematician with mastery over all the fields in his subject, which is one reason the institute Villani directs is named after him. Nowadays, said Villani, it is only possible for any one person to understand about a third of mathematics, and then only in a general sense. In depth no one can master more than a few per cent. As maths expands, its towers of knowledge grow ever higher and ever wider, which means that one specializes early in one's career. As a result, maths has become overwhelmingly a collaborative discipline. The cliché of the eccentric loner is no longer true, if it ever was. 'Maths is often at the interface, and at the interface it is better to take two specialists, one from either side. I was particularly good at exploiting collaborations with other people and finding resonances.' Villani believes that mathematics is currently going through a rich process of cross-fertilization. 'At the beginning you have one breed, then it separates, this one specializes, you get various ethnicities, and so on. Then they cross again. And it is interesting to have the cross after the specialization. We are at a time when different mathematical fields are coming together, and they are also coming together with other fields of science in a way that is much better than it was before.'

Henri Poincaré is perhaps best remembered for a statement he made in 1904 about a topological property of the sphere, which became known as the Poincaré conjecture. (And which is too complicated to explain in lay terms in a sentence, or even in several sentences.) For almost a century the conjecture was one of the most famous unsolved problems in maths, until in 2002 a 36-year-old Russian uploaded a proof onto an internet archive. By the time his peers had verified the proof, four years later, Grigori Perelman had abandoned mathematics. He was a recluse, living with his mother on the outskirts of Saint Petersburg (and resuscitating the cliché of the eccentric loner). In 2006 Perelman stunned the mathematical community by turning down a Fields Medal, saying that he needed no recognition beyond people understanding that the proof was correct. It was the greatest controversy since the award was first given in 1936. In 2010 the Clay Mathematics Institute awarded Perelman a $1 million prize for his Poincaré proof, but he refused that, too. Perelman's unclaimed award, a glass plaque on a stone plinth, now sits on the shelf of Villani's office, because the prize money was reassigned to fund a new chair at the Institut Henri Poincaré.

'Perelman is such a mystery,' agreed Villani. I asked if he had read Perelman's proof. 'With a bit of work I could read it. It's not so far from my area,' he replied. 'In mathematics people think that because we have [a proof] we should be able to decide on the spot whether it is true or false. But it is not [like that].' To get into Perelman's head, he said, would take a long time.

Perelman is one of six Russians to have won a Fields Medal since 1994, a period in which Russia claims more medallists than any other nation. France is the runner-up with five. However, if you include the Belgian who worked in France, and the Vietnamese and Russian who both hold French citizenship, then France leads with eight medallists, out of a total of 18. What's more, all of the French medallists work in Paris. The city has more professional mathematicians than any other. 'About a thousand [live here],' said Villani. 'A crazy amount!' One reason for France's tally of medals is their elite education system: all but one of French Fields medallists were undergraduates at the ultra-competitive École Normale Supérieure, which takes only 41 or 42 maths students a year. But history also plays a part. Fermat's Last Theorem, Cartesian coordinates, Pascal's triangle, Fourier transforms: the study of mathematics is a roll call of Frenchmen, and a matter of national pride. No Fields medallist is as much of a public figure in his own country as Villani is in France.

Villani was recently having a discussion with some physicists about Nicolas Carnot (1796–1832), who was the first person to explain how the steam engine worked theoretically. 'Carnot did not want for a second to build a machine. He didn't care at all,' exclaimed Villani. 'Yes, he was French! The English want to *build* the machine but the French want to *understand* it on a theoretical level. It has been so for hundreds of years!' And so it will continue to be. *Vive la différence.*

Integration is the part of calculus that concerns area, so when in 1876 the Scottish engineer James Thomson devised a machine to measure area he called it the 'integrator'. His contraption was an improved version of a nineteenth-century scientific instrument called the 'planimeter', commonly used by surveyors to calculate the areas of irregularly shaped sections of map. The planimeter consisted of a pointer connected to a wheel-and-disc mechanism, such that when you traced the pointer around the perimeter of an area, the mechanism gave an accurate reading for the portion enclosed.

Thompson showed plans for his integrator to his younger brother William, later Lord Kelvin, who realized in a flash its potential to mechanize computation. Just as integration is a component in a differential equation, Kelvin saw that integrators could be used as components in a machine to solve differential equations. Kelvin put integrators to use straight away in his 'tidal harmonic analyser', a machine he had invented to calculate the timing of tides.

Building on Kelvin's insight that a chain of integrators can model a differential equation, in 1927 Vannevar Bush at MIT constructed a 'differential analyser', a computer designed purely to solve differential equations. The 100-ton behemoth contained eight mechanical integrators on a desk as wide as a room, and was one of the first working machines able to perform advanced mathematics, predating the first digital electronic computers by a decade.

The differential analyser was an 'analogue' computer, because its mechanical parts were analogous to interactions in the physical system it was modelling. Bush's machine provided the template for many analogue computers right up to the 1970s, when the digital age made them, together with the slide rule, obsolete.

We learnt earlier that calculus was the handmaiden of Isaac Newton's laws of motion and gravitation. His mathematical innovations enabled him to establish a coherent body of formulae that describe how the forces acting on an object determine its position, speed and acceleration. In *Principia* he introduced a new concept, centripetal force, which is the force 'by which bodies are drawn or impelled, or in any way tend, towards a point as to a centre'. It is the force that makes objects travel in circles. Imagine a tennis ball attached to a piece of string. Hold one end of the string above you and whirl it so the ball makes circles in the air. The string is pulling the ball towards the centre with centripetal force.

The formula for centripetal force is $\frac{mv^2}{r}$, where m is the mass of the object, v its speed and r the radius of the circle, as illustrated below. The speed at each instant is perpendicular to the string, and the force acts down the string towards the centre. Newton's primary concern in *Principia* was centripetal forces acting on planets. In the eighteenth century, however, the force was of concern to transport engineers.

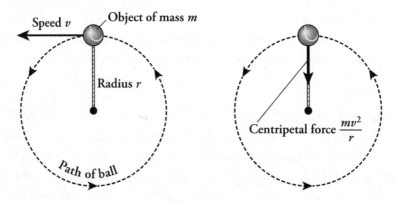

Centripetal force keeps the tennis ball moving in a circle.

The earliest railways used only straight or circular sections of track. This combination was problematic, since when a train moved from a straight to a circular section passengers felt a disagreeable lurch. A train travelling at a constant speed down a straight track has no forces acting on it. But the moment it meets the circular track, it is subject to a centripetal force. The force acts inwards, which gives the feeling of being jolted outwards. (Passengers are not being *moved* outwards. They are being diverted from a straight trajectory into a circular one, which, because the frame of reference in the carriage stays the same, gives the illusion of a force pushing outwards.)

'After half a century of railroading we are still using in our tracks only straight lines and circles,' wrote the American engineer Ellis Holbrook in 1880. 'Railroad men … seem to accept this barbarous combination as final, many without even asking what is the matter with it.' Holbrook's solution was a transition curve between straight and circular sections, along which a train travelling at constant speed would be subject to a centripetal force that increased linearly over time. Since the force is $\frac{mv^2}{r}$, and m and v are constant, Holbrook's transition curve required the value $\frac{1}{r}$ to increase linearly over time.

Before we get to Holbrook's curve, let's look more closely at the concept $\frac{1}{r}$. Mathematicians call this value the 'curvature' of a circle with radius r, and it is the measure of how much the circle deviates from a straight line. Below left are two circles, a small one with radius r and a large one with radius R, touching the dashed line at a point. The curvature of the small circle is *larger* than the curvature of the bigger circle, because it deviates further from the line. An intuitive way of thinking about the curvature of a circle is to see it as a measure of 'tightness' – the smaller the radius of a circle, the more tightly curved it is, and the greater the curvature.

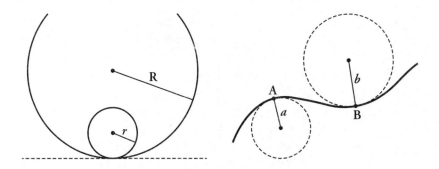

The smaller the radius of a circle, the greater its curvature.

The curvature of a circle with radius r is $\frac{1}{r}$ at every point. The curvature of the curve like the one above right, however, changes continuously as you move along it. To calculate the curvature at any point, we consider the 'best-fit' circle, which is the circle that touches the curve and approximates it at that point. I have drawn the best-fit circles at A and B. Since the radiuses of the circles are a and b, the curvature at A is $\frac{1}{a}$ and at B is $\frac{1}{b}$. An intuitive way of thinking about the best-fit circle is to imagine that the curve is a road. You are driving along it and the steering wheel jams at, say, A. If you keep on driving, the car will move round the best-fit circle at A.

Holbrook's idea for a transition curve was therefore for a piece of track whose curvature increases linearly as you move along it, since this is the curve for which centripetal force increases linearly. He may or may not have known that he was, in fact, describing a famous curve first investigated by Leonhard Euler in the eighteenth century, the clothoid, illustrated below. At point zero the curve is a straight line, which has zero curvature, and as you move away from zero (in both directions) the curve gets steadily tighter; its curvature increases in proportion to the distance travelled. The result is a spiral coiling in ever-decreasing circles until it converges on two dots. This shape is named after Clotho, the youngest of the three Fates, who spun the threads of life as the curve spins around its two poles.

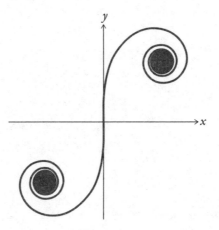

The clothoid.

From the end of the nineteenth century, the clothoid became the standard railway transition curve. Or rather, the central section of the clothoid did. Imagine that a piece of straight track joins the curve at 0 and then travels along it. The curvature slowly increases until equal to the curvature of a subsequent piece of circular track. When, in the twentieth century, the car replaced the train as the predominant method of motorized transport, the clothoid became a staple of road design for the same reasons. It is the most comfortable curve to drive along between straight and curved sections of road. The motorway network is a living museum of clothoids. The distinctive shapes are still used for bends in the route, in slip roads and, most plentifully, in spaghetti junctions, with many transitions between straight and circular sections. If you were an alien flying low across a countryside of motorways and rail lines, you might well conclude that the clothoid is human civilization's favourite curve.

The clothoid also solved a problem in fairground design: what is the safest shape for a roller-coaster loop-the-loop? In the mid-nineteenth century a Parisian engineer, M. Clavières, designed a fairground ride in which a single car raced down a straight track and somersaulted round a thirteen-foot-high circular loop before rolling up a smaller straight track to an end station. Several of these 'aerial railways' were built in France but they were all closed down because they caused so many neck injuries when the car went from straight to circular track, and for more than a century fairground promoters assumed that safe loops were impossible. It was only when the German engineer Werner Stengel looked at the problem in the early 1970s that he realized that the clothoid was the answer. Stengel designed the first loop-the-loop of the modern era, Great American Revolution, which launched at Six Flags Magic Mountain in California in 1976. The car descends a lightly sloping straight piece of track, and enters a clothoid until the curve has a

Le loop-the-loop, Le Havre, 1846.

radius of 7m, which is just when it begins to double back, as shown below. The car stays on the 7m-radius circle for just under half a revolution, at which point a mirror version of the first clothoid takes the car from circular back to straight. 'The transition is very soft,' said Stengel. 'A change in force makes a good roller coaster, but the change should be acceptable for the body.'

Great American Revolution was an immediate success, gaining a particularly seventies-style tribute: it was the subject of the disaster movie *Rollercoaster*, in which villains planned to bomb it on its launch day. Since then about 200 looping roller-coasters have been built around the world, all of them using Stengel's principle. The inverted teardrop of the clothoid roller-coaster is both a modern symbol of our insatiable appetite for thrills and a monument to the mathematics of Isaac Newton. It is the mechanical curve reincarnated as a dazzling monster of steel.

Newton's physical laws flowered from the tiny seed that was the infinitesimal, that quantity smaller than anything else but larger than zero. Despite its fecundity for producing new science, however,

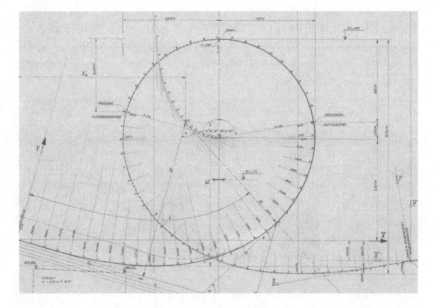

Werner Stengel's original drawing of the Great American Revolution.

the infinitesimal was also ridiculed for being self-contradictory. 'What are these … evanescent increments?' jibed the philosopher Bishop George Berkeley a few years after Newton died. 'They are neither finite quantities nor quantities infinitely small, nor yet nothing. May we not call them the ghosts of departed quantities?' Berkeley's stinging remarks caused uproar among scientists, who saw calculus, quite rightly, as the greatest mathematical advance of the Enlightenment. Yet the cleric had a point. The infinitesimal was not a rigorously thought-out idea, even if it did produce the right answers. The debate he provoked set mathematics on a path of soul-searching and self-reflection. Which concepts are allowed and which are not allowed? How much sense does maths need to make?

The Titl of This Chapter Contains Three Erors

Here's a teaser. One day I climb up a mountain, I sleep on the summit, and on the following day I walk back down the same path. Am I ever at the same altitude at the same time on different days?

Think about it for a second.

Or two.

I love this problem because it is simple to state and has a simple solution.

Now turn the page.

The answer is yes. Imagine both journeys are taking place on the same day. I'm simultaneously walking up from the bottom, and down from the top. It is inescapable that at one point I will bump into myself, and, when I do, time and altitude coincide.

If you accept this argument that there must be a time on both days when I am at the same altitude, then I am happy. My proof worked. Mathematical proof is simply a device that one person uses to convince another person that a mathematical statement is true, and I have convinced you.

A more fastidious mathematician, however, may not accept my argument. He or she may reject it for being insufficiently rigorous. Where is the proof that I will bump into myself? Let's draw a diagram that plots my ascent from the base of the mountain at altitude A to the summit at altitude B, and superimposes my descent the following day, as illustrated below. The question now becomes: is there a point where the two lines cross? Most readers will reply that of course there is! The fastidious mathematician will remain unconvinced.

Until the end of the eighteenth century it was assumed that if a curve like the one in the diagram goes from altitude A to altitude

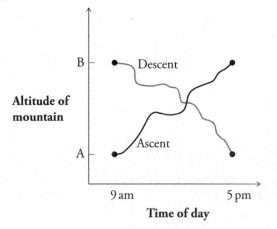

The Englishman who went up a hill, and then came down again.

B, then it must pass every single altitude between A and B. This statement seems intuitively self-evident. Indeed, it was part of what defined a continuous curve. But as mathematicians looked more carefully at properties like continuity they concluded that clearer definitions were needed. Statements that had been taken for granted were reclassified as theorems that required proofs from even more basic assumptions. The statement above, that a continuous curve with minimum value A and maximum value B must pass through every intermediate value, was one of them, and it is now called the 'intermediate value theorem'. Its proof is complicated, and taught only at university level, but it will be enough to convince our fastidious friend. Consequently, he or she will accept that the two curves in the diagram cross at a point, since this statement follows from the proof in a few steps.

Experiments drive science. Proofs drive maths. There are lots of ways to conduct experiments, as there are lots of methods of proof. We will look at some of them in the following pages. We will also examine how attitudes towards proof have changed, and speak to an anonymous member of a secret society dedicated to mathematical rigour. But first: let's eat.

The intermediate value theorem may seem obvious, but it quickly leads to some interesting results. One corollary is the pancake theorem, which I prefer to describe in more savoury terms. If you spill salt on a table (or serve up a pancake) we can prove that there is a line which splits the salt (or pancake) into two sections of equal area, at any angle you like. The method is shown overleaf. First, draw a line outside the salt at your preferred angle, labelled X, and then move it towards the salt while keeping it parallel to the starting position. The line starts to cross the salt at A, when no area is covered, and leaves it at B, when the whole area has been crossed. The area of salt covered changes continuously as the line crosses from A to B. According to the intermediate value theorem, the line must pass a position where the area covered is exactly half the total. Our proof does not tell us where the cut is, only that such a cut must exist.

Now say you spill salt *and* pepper on a table. We can also prove

235

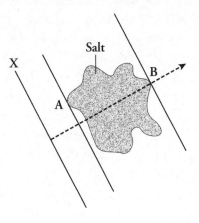

The salt theorem.

that there is a line which divides both salt and pepper into two sections of equal area. Start by defining a line X that cuts the salt in half and does not touch the pepper, as illustrated opposite. Rotate the line clockwise, always making sure that it maintains the division of the salt into two equal parts. We know we can do this because, from above, there is a bisection of the salt at every angle. Our rotating line touches the pepper at A and leaves it at B. The area of pepper covered grows continuously from zero to the maximum, so the line must pass a point where it also divides the pepper in two. In the illustration the areas of salt and pepper are separate, but even if they overlap there is always a line that divides both into two equal areas.

Between the First and Second World Wars, a clique of mathematicians in Lwów, Poland, met regularly in a coffee shop, the Scottish Café, to discuss mathematical morsels such as the pancake theorem. Hugo Steinhaus, a principal member of the group, wondered whether the theorem could be extended into three dimensions. 'Can we place a piece of ham under a meat cutter so that meat, bone and fat are cut in halves?' he asked. His friend Stefan Banach proved that such a cut is possible, using a theorem attributed to two others in the group,

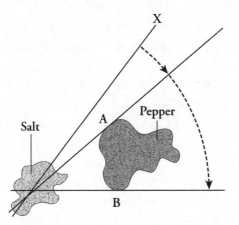

The salt and pepper theorem.

Stanislaw Ulam and Karol Borsuk. Banach's result has subsequently been popularized as the 'ham sandwich theorem', because it is equivalent to stating that one can divide a ham sandwich in two with a single slice that cuts each slice of bread and the ham into two equal sizes, no matter how each piece is positioned and whatever its shape.

The mathematicians who gathered in the Scottish Café kept a thick notebook of all the questions they asked each other, which they entrusted to the care of the head waiter when they went home. Eventually known as *The Scottish Book*, it is a unique collaborative work, and not just because of how it was written. (It was never published as a book, but some of its problems appeared later in journals.) Steinhaus, Banach and Ulam were all significant mathematicians, as talented a trio as has ever emerged in any one place at any one time. A few days after Steinhaus wrote what turned out to be the last problem in *The Scottish Book*, in 1941, the Germans entered Lwów and Steinhaus, who was Jewish, went into hiding, surviving the war in a small town near Krakow using the identity of a deceased forest ranger. During these years he reconstructed from memory most of the maths he knew, and worked on new problems, including another one inspired by food.

Steinhaus wanted to know the fairest way to divide a cake between people who want as much as they can get. When there are only two people, the approach since ancient times has been that one cuts and the other chooses. This motivates the cutter to be as accurate as possible, since if there is a noticeable difference between the portions he will be left with the smaller one. Steinhaus was the first person to work out how *three* people can divide a cake fairly. (A description is in Appendix Seven on p. 300.) Since Steinhaus the mathematics of cake cutting has become a large field, with applications in economics and politics. Many variations of the problem exist, depending on how many people there are who want portions, and how they value the different parts of the cake. One ingenious method invented in the 1960s, which can be used for any number of people, concerns a moving knife. The knife is positioned at the side of the cake and then moves very slowly across it. When someone shouts 'STOP!' the knife slices at that position. The person who shouted out receives the slice. The knife then continues for the remaining participants.

Steinhaus is remembered for perhaps the two most widespread food metaphors in maths – the ham sandwich theorem, and the fair division of cakes. He was constantly thinking of food. Tragically, for much of his life, he had little of it.

A common method of proof is *proof by contradiction*, in which a statement is shown to be true because by being false it implies a contradiction. For example:

Theorem: *All numbers are interesting.*
Proof: Let's assume that the theorem is false; that there are some boring numbers. If this were so, there would be a smallest boring number. But the 'smallest boring number' is, by that very fact, an interesting number. In other words, the term 'smallest boring number' contradicts itself. We have our contradiction. The theorem cannot be false, and so must be true.

The Greek thinker Aristotle was one of the first to study the nature of proof. He developed a system of logical reasoning concerned with whether the truth of premises necessitates the truth of conclusions. His concern was philosophy, yet the idea that truth can flow from premises to conclusions through logical deduction was to have a more profound influence on mathematics. Indeed, since the Greeks, maths has been the study of exactly that: how true premises lead to true conclusions via proofs.

In the third century BCE, Euclid wrote *Elements*, his landmark treatise on geometry. It had a very distinctive literary style and a revolutionary conceptual framework. Euclid started with a small set of assumed truths, the axioms, and from them deduced all the other truths, the theorems. His way of systematizing knowledge is called the 'axiomatic method'.

Elements reads like a recipe book for aspiring geometers. It begins with a list of ingredients: definitions for 26 terms, and ten assumptions we are allowed to take as true, such as the assumption that given two points we can draw a line between them. Euclid then takes us through the dishes he wants to cook, which are the 'theorems', together with step-by-step instructions, which are the 'proofs'. The first theorem is how to construct 'an equilateral triangle on a given finite straight line'. The second is how 'to place a straight line equal to a given straight line with one end at a given point'. In each proof Euclid only uses the assumptions he listed at the start of the book, and each step follows logically from the step before. The style of declaring one's assumptions, and then slowly building up knowledge through theorems and proofs, became the template for all subsequent mathematical works.

One of the best-known theorems in *Elements* uses proof by contradiction.

Theorem: *There are infinitely many prime numbers.*
Proof: First, a warning. A proof cannot be read as fluently as prose. It is perfectly normal to read one over several times before it sinks in. And second, let's be clear about what Euclid is trying to do. The prime numbers (2, 3, 5, 7, 11, 13 …) are those numbers greater than 1 that are only divisible by themselves and 1. Euclid will show that if the theorem is false we get a contradiction. More specifically, he will show that if there is only

a fixed number of prime numbers, it is possible to generate a new prime number, thereby contradicting the statement that there is only a fixed number of them. The theorem cannot be false, so it must be true.

Step 1 Let a, b, c ... k be our fixed number of primes.

Step 2 Multiply every number in this set together, to create a number $a \times b \times c \times ... \times k$. Let's call this number M.

Step 3 Increase this number by one, to M + 1.

Step 4 Is M + 1 prime?

(i) If M + 1 is prime, then we have achieved our aim of finding a prime number not in the original set.

(ii) If M + 1 isn't prime, there must a prime number p that divides it. Either p is one of our original primes or it isn't. If it isn't, we have our new prime. If it is, we know that p divides M, since all the original primes divide M. But now we have a situation where p divides both M and M + 1, which is impossible, since the only number that divides two numbers spaced one apart is 1, which is not prime.

In conclusion, either M + 1 is a new prime, or M +1 is divisible by a new prime. Either way, Euclid's job is done. He has proved that the finite set does not cover all primes.

Euclid's proof uses a tactic called *reductio ad absurdum,* in which an absurd conclusion demonstrates the falsity of the premise. In step 4 (ii), the absurd conclusion is that p must divide both M and M + 1, and the false premise is that p is in the finite set of primes. In *A Mathematician's Apology,* the Cambridge don G. H. Hardy wrote that Euclid's proof is 'as fresh and significant as when it was discovered – two thousand years have not written a wrinkle on [it].' The proof is short and concise, requiring no concepts beyond addition, multiplication and division. '*Reductio ad absurdum,* which Euclid loved so much, is one of a mathematician's finest weapons,' Hardy added. 'It is a far finer gambit than any chess gambit. A chess player may offer the sacrifice of a pawn or even a piece, but a mathematician offers *the game.*'

Reductio is also one of a comedian's finest weapons. The use of irony to push for absurder and absurder conclusions, thus displaying the premise as increasingly ridiculous, is otherwise known as satire.

In fact, I find Euclid's proof of the infinitude of primes inherently comic. To find a new prime number, Euclid must first create a number, M, that is not only ridiculously big but is also exactly the opposite of what he is looking for, since it is divisible by every known prime. Then, by adding 1, the smallest possible number, he turns the situation upside down. The wafer-thin extra element undermines the massive, mega-divisible monster M and its constituent primes by savagely exposing their limitations. Like the sarcastic catchphrase popularized by the film *Wayne's World*, Euclid is saying: 'This group of primes covers every number … *not!'*

Maths is full of jokers.

From the day we humans are able to hold a pen we take pleasure in doodling. A common genre is to randomly criss-cross a section of paper with lines, and to start shading the regions created. This style is particularly satisfying since the doodle can always be covered in such a way that shaded regions only share sides with unshaded regions – and vice versa, that unshaded regions only share sides with shaded ones. This type of shading is called 'two-colouring', since the entire pattern contains only two colours. To show why we can always two-colour a doodle, we need to introduce another common mathematical tool: *proof by induction.*

In philosophy and empirical science, induction is the principle that if an event has been observed many, many times in the past we can assume that it will happen again in the future. The sun, for example, has risen every morning for as long as we have known. It is reasonable to assume that it will rise tomorrow. We cannot prove that the sun will rise tomorrow, but we can be confident. In mathematics, on the other hand, we cannot assume there is a pattern just by looking at past instances.

Consider the five circles below. The first has a single dot on the circumference, the second has two, the third three, the fourth four and the fifth five. Connect the dots with lines, and count the regions in each circle. The circles can be divided into 1, 2, 4, 8 and 16 regions. The pattern is striking. It's the doubling sequence! Surely we can surmise that if we connected the dots between six points on the edge of a circle, the number of regions would be 32?

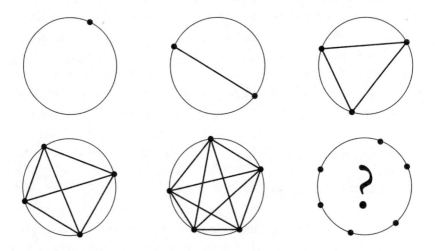

Count the regions in each circle and guess what comes next.

A thunderous NO! There are 31 regions with six dots, and as we add more dots, the sequence continues 57, 99, 163, 256, 386 ... There is a pattern, but it is not the doubling sequence. One should never draw a conclusion from a limited number of observations, no matter how promising it seems.

In maths, proof by induction is a method that lets us know when a pattern will continue for ever. If we have a sequence of statements such that:

(a) The first statement is true.

    AND

(b) If the nth statement is true then the $(n + 1)$th is true.

Then we can infer that *all* the statements are true.

Proof by induction is analogous to the toppling of dominoes. If dominoes are stacked in a line such that if the nth domino falls, it will push the $(n + 1)$th domino over, then for every domino to fall we only need to topple the first one.

Returning to our original objective, to show that doodles can be two-coloured, we need to show that:

(a) A doodle with 1 line can be two-coloured.
(b) If a doodle with n lines can be two-coloured then a doodle with $n + 1$ lines can be two-coloured.

Showing that statement (a) is true is trivial: draw a line across a page and shade one side. Showing that statement (b) is true demands more thought.

We embark on the proof by considering a doodle with $n + 1$ lines, as shown below in (i). (Obviously, for the sake of an illustration, I have to choose a value for n, so we need to be careful to make sure that the following proof applies to the general case for any n.)

When we remove a line, we have a doodle with n lines, shown in (ii). Assume that the doodle with n lines is two-colourable, (iii).

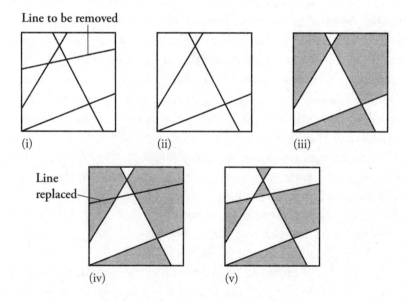

Proving the two-colour doodle theorem by induction.

Now reinstate the line we removed, (iv). On one side of the line, flip the colours, so white becomes shaded and shaded becomes white. Each region above the line is now adjacent to a section below the line with a different colour. We now have a doodle with $n + 1$ lines that is two-colourable (v).

In other words, we have demonstrated that statement (b) is true. Our inductive proof is complete: all doodles can be two-coloured. (This proof only concerns doodles made from lines criss-crossing a square. It is also the case that any 'squiggle' doodle, where the pen starts and ends at the same point but whose path can involve any type of elaborate looping, spiralling and intersecting, is two-colourable. That proof is a little trickier.)

Elements became the most influential text in mathematical history not because of what it did – reveal facts about prime numbers, triangles, and so on – but because of how it did it. The beauty in its pages is its rigour. Euclid is meticulous. He takes no shortcuts, he offers no opinions and he asserts nothing he cannot prove. If you accept that Euclid's ten assumptions are true, then you must accept that all of the book's 465 theorems are true, and will always be true. *Elements* is a showcase of the axiomatic method, a testament to the power of deductive thought.

It is said that *Elements* has appeared in more editions over more centuries than any other book apart from the Bible. The comparison is fitting. For almost two millenia, *Elements* was treated as if it was a sacred text, the axiomatic method accepted as dogma. In the seventeenth century, however, the first rumblings of impiety emerged. Euclid relied on his axioms and definitions being as self-evident as possible, and certainly not self-contradictory. Yet as we saw in the previous chapter, the infinitesimal, a quantity that is both something and nothing, was exactly that. Newton and his contemporaries employed the infinitesimal because it permitted a blossoming of new theorems, even though they had to close their eyes to the Euclidian heresy it involved.

Eventually, though, mathematicians realized that to be sure calculus was free from contradictions it needed to have a sounder footing. The solution was to base calculus not on the infinitesimal but on the more solid concept of a limit. Assumptions were simplified, definitions were clarified and a new field was born – 'analysis', which is the term now used to cover all subjects connected with calculus, continuity and infinite processes. One signature achievement of early analysis was the intermediate value theorem, mentioned at the beginning of this chapter, which states that a continuous curve will cover all values between its minimum and its maximum.

The nineteenth-century interest in rigour was reflected in fields other than analysis. Euclidean geometry, for example. A German mathematician, Moritz Pasch, looked closely at *Elements* and made an amazing discovery: there were holes in Euclid's reasoning that no one had spotted, despite it having been the most studied maths textbook of all time. Euclid takes it for granted that if three distinct points are on a line then one of them is between the other two – an observation that again is completely obvious, but if we are going to hold Euclid up to his own standards, one that he should have declared as an axiom. Euclid had made the sloppy mistake of letting his eyes influence his deductive process. A new, watertight Euclidean system was proposed by David Hilbert in 1899 with twenty-one assumptions.

Numbers were also placed under a new type of scrutiny. They are the core of all mathematics, indeed of all science. But what exactly *is* a number and why is it that 1 + 1 = 2?

In 1879, the German mathematician Gottlob Frege published *Begriffsschrift*, or 'Concept-script'. The book introduced an elaborate calculus, with its own notation, for expressing the truth and falsity of statements. It was the birth of 'mathematical logic', the use of mathematical reasoning to investigate mathematical reasoning.

Frege wanted to give a rigorous answer to the question 'what is a number?' For this challenge he borrowed another concept, the set, from his contemporary Georg Cantor. Often in mathematics, a simple-sounding word means something complicated. Not with the set. A set is just a group of things, a collection of stuff united by a common property. A set can be a box of apples, a peloton of cyclists or a galaxy of stars. The number two, wrote Frege, is the set of all sets with two things in them.

Frege devised a system in which numbers are defined as sets, axioms are written using his concept-script, and the laws of arithmetic can be proved to be true. His idea was to reduce arithmetic to a watertight framework of logical operations based on uncontroversial assumptions, such as, for example, 'the negation of the negation of statement A implies statement A'. Since numbers and addition are easy concepts to master, you might think that Frege had a simple task. In fact, it was brain-stretchingly difficult. Unlike all previous mathematicians, who had used numbers and arithmetic as bricks to build the house of mathematics, Frege was digging down into the foundations.

Frege published his theory in *The Basic Laws of Arithmetic*. The first volume appeared in 1893. When the second volume was at the printers, however, Frege received a bombshell. Bertrand Russell, a philosophy fellow at Cambridge University, sent him a letter pointing out a contradiction. Since the aim of reducing arithmetic to logic was to provide a structure free from contradictions, there was nothing more devastating than to find one. Frege hastily wrote an addendum: 'A scientist can hardly meet with anything more undesirable than to have the foundations give way just as the work is finished.' His use of the word 'undesirable' has been called the greatest understatement in the history of mathematics.

Russell had uncovered the curse of self-reference.

Some of my favourite one-liners are self-referential statements:

i should begin with a capital letter.

This is to be or actually not two sentences to be, that is the question, combined.

This sentence is a !!! premature punctuator

The granddaddy of all self-referential sentences, however, is credited to Epimenides from Crete, who said 'All Cretans are liars.'

Epimenides is not just referring to himself but also contradicting himself. If he is telling the truth he is lying, and if he is lying he is telling the truth. His utterance – the 'liar's paradox' – has been reinterpreted many times. Answer the following question with a yes/no answer: 'Will the next word you speak be no?'

Bertrand Russell realized that a self-referential paradox severely wounds Frege's project, and maybe kills it outright. The beauty of using sets as a basis for arithmetic is that they are easy to understand: a set is simply a collection of things. Russell, however, devised the following set:

The set of all sets that do not contain themselves.

Most sets do not contain themselves. The set of shoes is not itself a shoe. But some sets do. The set of concepts, for example, *is* a concept. Now consider Russell's set. Does it contain itself? If we assume it does, we are led to the conclusion that it doesn't, and if we assume that it doesn't, we have to conclude that it does! The set implodes in self-contradiction. Russell provided the analogy of a barber in a village who has a sign on his wall: 'I shave only those men who do not shave themselves.' Who shaves the barber? If he shaves himself, he does not shave himself, and if he does not shave himself, he shaves himself. We are lost in an infinite self-contradictory loop.

Russell's paradox shows that sets as Frege imagined them are ill-equipped to provide a logically sound basis for arithmetic. Self-reference allows self-contradiction to contaminate the system. Yet rather than dismiss Frege's project as flawed, Russell became its biggest champion. The dream of putting maths on a solid logical foundation was too intoxicating to abandon, and for the next ten years, together with his colleague Alfred North Whitehead, Russell worked to fix the system. The men accepted Frege's premise that the set could provide a valid foundation for numbers. To banish the paradoxes of self-reference, however, they proposed a strict hierarchy of sets. On the first level are objects, such as books or cats. On the second level are sets of level-one objects, such as the books on my shelf or the cats on my street. On the third level are sets of level-two objects, such as the bookshelves of maths authors or London cats grouped per street. Russell's paradox cannot arise since a set can only be a member of one on a higher level, so can never be a member of itself.

Russell and Whitehead introduced notation, definitions and axioms that they spelled out with supreme diligence and care. Yet in order to be as simple and as clear as possible, they ended up writing one of the most complicated and unreadable major texts in the history of mathematics. Only on page 379 are they able to state that $1 + 1 = 2$. When they submitted *Principia Mathematica* for publication, the publisher turned it down because he could find no reader able to understand it. Russell and Whitehead paid to publish it themselves. Writing *Principia* was so mentally exhausting that Russell never wrote on mathematics or logic again.

The Polish logician Alfred Tarski proposed a hierarchy of language, much like Russell's hierarchy of sets, which resolves the liar paradox. There is a level 1 language, and a level 2 'metalanguage' that talks about statements in the level 1 language, and a level 3 meta-metalanguage that talks about statements in the level 2 metalanguage, and so on. The truth or falsity of a statement can only be discussed in the next meta-level up, so it is meaningless for a statement to ascribe truth or falsity to itself. As Russell once explained, if Epimenides states that 'I am telling a lie of level *n*', he is telling a lie, but a lie of level $n + 1$.

Comedians, as well as logicians, rely on metalanguage. If a joke fails, it is often possible to salvage some humour with a joke about the joke.

Principia remains an unread doorstop. Nevertheless, its quest to provide an axiomatic foundation for an arithmetic free of paradoxes was enthusiastically taken up by others. Axiomatic set theory is now seen as one of the great intellectual triumphs of the early twentieth century, and it has led to wonderful work in maths, logic and philosophy. The standard system of axioms is called ZFC, after the mathematicians Ernst Zermelo and Abraham Fraenkel, and the 'axiom of choice'. The axiom of choice states that if you have an infinite number of sets, each containing at least one item, it is possible to create a new set that contains exactly one item from each set. The axiom seems reasonable enough, yet it was hugely controversial. One of the most passionate debates in set theory was whether or not to accept it in the system, because if you do, some very strange things start to happen.

Stefan Banach, the Pole who solved the ham sandwich theorem in the Scottish Café, and Alfred Tarski, the logician who proposed a Russellian hierarchy of language, showed that if you accept that the axiom of choice is true, then the following theorem is also true:

It is possible to divide a solid sphere into a finite number of pieces that you can reassemble in a different way to make two identical copies of the original sphere.

The theorem is better known as the Banach-Tarski paradox. The word 'paradox' is used because it appears to contradict the laws of physics, yet the proof contains no logical contradictions. Reassembly is physically impossible because the pieces are not continuous slices, but are each infinite scatterings of points. Still, the theorem is mind-blowing. It follows that any solid object can be cut up and reassembled into any other object, so a pea could be turned into the sun. (Despite such bizarre consequences, however, most mathematicians now accept the axiom of choice.)

If jokes are about unexpected conclusions, then the Banach-Tarski paradox is the funniest theorem in maths.

In the late seventies, when I was about eight years old, the focus of maths class stopped being numbers and became sets. I remember it well. An oval with some dots in it was one set, another oval with some dots in it was another set. We would connect the dots in one set to the dots in another set, which would tell us which set had the most dots. I never saw the point of these exercises, and I don't think my teachers did either. After about a year sets disappeared from class and the next time I encountered them was in my second year at university. If you went to school in the 1960s, 1970s or 1980s you may also have briefly been exposed to set theory. Its presence in the curriculum can be traced back to Nicolas Bourbaki, the most prolific mathematician of the twentieth century.

In 1939 Bourbaki published the first book of an ambitious series called *Éléments de Mathématique*. 'Whereas in the past it was thought that every branch of mathematics depended on its own particular intuitions which provided its concepts and prime truths,' he wrote, 'nowadays it is known to be possible, logically speaking, to derive practically the whole of human mathematics from a single source, the theory of sets.' The title was a nod to Euclid. Just as *Elements* formalized Greek mathematical knowledge within an axiomatic system, based on the properties of points and lines, so *Éléments* aimed to formalize modern mathematical knowledge within an axiomatic system based on the properties of sets. The choice of the word *mathématique*, 'mathematic', in the singular, emphasized Bourbaki's belief in the unity of the subject. *Éléments* contains dozens of books, not just on set theory, but also on subjects as varied as algebra, analysis and topology. The series runs to more than 7000 pages and is one of the most influential scientific texts of the twentieth century. Bourbaki also had a distinguishing characteristic that made him unique among his contemporaries. He did not exist.

In the early 1930s, a group of young French mathematicians decided that their university textbooks were out of date, and that collectively they would write new ones. They decided on the nom de plume Nicolas Bourbaki, after Charles Denis Bourbaki, a French general who in 1862 had declined the throne of Greece, and then, after suffering a humiliating defeat in the Franco-Prussian war, tried to shoot himself and missed. Nicolas Bourbaki, they said, was from Poldavia, a country mentioned in the Tintin adventure *The Blue Lotus*. The group adopted a code of secrecy and made resignation compulsory at the age of fifty. Like the Polish mathematicians who gathered at the Scottish Café in Lwów at around the same time, the Bourbaki group enjoyed mixing fun with mathematics. During the first of their regular congresses in the countryside, a few members went to a local lake and jumped in naked, shouting 'Bourbaki!'

Bourbaki maths, however, was deadly serious. The group devised a writing method that meant each of the books took years to write. After long discussions about the content of each volume, one member would compose a draft. At a subsequent congress, the draft would be read out line by line, and each line had to be approved by every member. The style was also unique. The aim of the series was to deduce everything from first principles, without any recourse to physical or geometrical intuition. Illustrations were not used, as it was thought they could be misleading. 'Rigor is to the mathematician what mortality is to men,' said André Weil, one of the group's founder members. There were no analogies, digressions, omissions, sketches or exercises for the reader. Such was the insistence on axiomatic purity that it takes the first book two hundred pages before the number 1 is defined, and then only in abbreviated form. (In its fully expanded form, the book says, the number 1 would require many tens of thousands of symbols. In 1999, the British set theorist A. R. D. Mathias claimed that the Bourbaki method would actually require 4,523,659,424,929 symbols and 1,179,618,517,981 links between them.)

The series was meticulously structured. Each book was only allowed to refer to material in previous books, and could not refer to anything in books by other authors, thereby constructing a gigantic logical edifice based on set theory. Even though the members of the group were young, they were all accomplished mathematicians, who also published work independently. André Weil, the brother of the philosopher and activist Simone Weil, was perhaps the most gifted. In 1939, the year the first book of *Éléments* appeared, war broke out and Weil fled to Finland. Police raided his Helsinki apartment and deported him on suspicion of espionage, after finding a letter in Russian (describing maths) and a packet of calling cards belonging to Nicolas Bourbaki, of the Royal Academy of Poldavia. Back in France, Weil was jailed for failing to report for army duty. He liked being in prison. 'My mathematical work is proceeding beyond my wildest hopes, and I am even a bit worried,' he wrote to his wife. 'If it is only in prison that I work so well, will I have to arrange to spend two or three months locked up every year?'

The second book of *Éléments* appeared in 1940, and the third in 1942. After a hiatus due to the war, more volumes appeared at the end of the decade. A new group was drafted in as the old one reached the age limit, and the books kept on coming. By the fifties the Bourbaki library dominated university mathematics in France, and would do so for the next two decades. The sect began to resemble a mafia, as members and ex-members – including many of France's most brilliant mathematicians – entered top university positions. In translation, Bourbaki also had considerable influence in the English-speaking world.

The heyday of Bourbaki coincided with an escalation in the Cold War. Western governments realized they needed to overhaul their science education to keep up with the Soviets, who had just launched Sputnik, the first satellite. Bourbachiste ideology, that abstract formal structures were more important than intuition and problem-solving, filtered down from universities to schools. Politicians and educators decided that the answer to the Red Menace was a curriculum based on set theory. Maths teaching was reformed, and in the sixties and seventies a generation of children were introduced to the 'new maths': the joy of sets.

The direct influence of Bourbaki in university lecture halls and school classrooms has waned. Areas of research like fractals, for example, which are reliant on computers and visual images, have made Bourbaki's preoccupation with structure appear passé. In the last few decades, maths has advanced by interacting with other sciences, not by isolating itself from them. As a consequence, children are no longer taught set theory. But contrary to reports of his demise, Nicolas Bourbaki, now approaching his eightieth birthday, remains alive and well.

Five mathematicians now form the core of the group. I met one of them in a café across from the Luxembourg Gardens in Paris. The code of secrecy still holds and all I am allowed to report is that he has a beard, and was wearing a purple shirt and a straw hat. He is also an eminent professor. I asked how many people knew of his role in Bourbaki. 'A lot of colleagues know perfectly well, but I would not acknowledge it. There is a lot of resentment,' he said. 'Some people say Bourbaki has been useless, that it should stop.'

The most recent book in the *Éléments* series, on algebra, was published in 2012, and a new one, on topology, is being prepared. The case against Bourbaki is that its obsession with rigour ultimately did French mathematics a disservice. The books are hard, and therefore poor educational tools, and leave no space for creativity and intuition. 'Even close colleagues believe they are not the books that current mathematicians need,' the purple-shirt said. Does he think they are, I asked? 'The answer is not clear. But what *is* clear is that in this kind of work – where we all get together, read together line by line, and everybody takes the time to contradict and to correct – something unusual comes out, and hopefully something good. The ideas in the book are a mixture from a lot of people. Mathematicians cannot do everything by themselves.'

I asked whether he thought Bourbachiste levels of rigour were outdated? 'I think that rigour is more relevant than ever,' he replied, adding: 'There is a difference between rigour and dryness. We try to be rigorous but not dry.' In fact, he believes that modern university textbooks all owe something to Bourbaki. 'Now it is standard practice to admit that your proof is not rigorous, for the level of the book. In some way the level of rigour practised by mathematicians is [the Bourbaki] one.' One criticism he does accept, however, is of the first book. 'Some Bourbaki books are good. Some are extremely good. But the set theory book is crap.' He cringes when I remind him of how Bourbaki defines 1. 'That part is not good. You don't need to know what 1 is. You need to know what you can do with 1.'

My interviewee told me, however, that he is immensely proud to be in Bourbaki. He was thirty, and recently appointed a professor, when he received his first diktat: Nicholas Bourbaki ordered him to be present at the next congress, at a chateau in the Loire. When invited, most people accept, he said, although the few women who have been invited have all said no. Now a fully-fledged member of the group, he feels a historic duty to help it finish the work it set out to do, which is to complete the *Éléments* series. The final four books are planned. He knows they are unlikely to appear before he reaches fifty and resigns. The age limit is a good thing, he said, because it keeps the group alive.

———

Set theory is one approach to constructing a foundation for mathematics. Another is currently under way, using computers. A 'proof assistant' is a piece of software that checks whether the logical inferences in a proof are correct. The aspiration is that one day computers will prove all of mathematics. If you want to know whether a theorem is true or not, you'll press a button.

The first major theorem to be proved with help from a computer was the four-colour map theorem. Earlier we saw that all doodles can be 'two-coloured', meaning that we can shade all doodles in such a way that two adjacent regions are never the same colour. In 1852 Francis Guthrie, a South African in London, was colouring in a map of the counties of England. He noticed that four colours were needed to ensure that no county shared a border with a county of the same colour. Experimentation seemed to show that four colours were sufficient to colour all maps in this way. But for more than a hundred years no one was able to prove it, until, in 1976, Kenneth Appel and Wolfgang Haken at the University of Illinois did just that, using a supercomputer to check through all possible map configurations. Mathematicians responded uneasily. In principle, one should be able to check every line of a proof. But the computer performed too many calculations for checking to be possible, contravening the paradigm for proofs that had existed since Euclid. As well as philosophical objections to the proof, however, there were other more practical worries. Programs are full of bugs. How could Appel and Haken be sure there were no bugs in their software? They couldn't, and, in fact, computer errors continue to be found in their proof, although these have all been fixed. In 1995 a team at Princeton University produced a more streamlined computer proof of the four-colour map theorem. And in 2004 Georges Gonthier, at Microsoft Research Cambridge in England, checked it with a proof assistant, although he had to translate all the concepts into a special programming language that the assistant could understand. The question then arises: how can you make sure there are no bugs in the assistant? You can't completely, but you can have a higher degree of certainty than in the original proofs, since the assistant has already been stress-tested on many other jobs. The four-colour map theorem is now one of the most thoroughly verified proofs in maths.

After an initial resistance to computer-aided proofs, most mathematicians now accept them. Some even dream that one day every theorem will be translated into a universal proof-checking computer language, creating a giant formalistic system that contains all provable mathematical knowledge, and in which every statement is rigorously derived from a set of basic lines of code. When that happens, we should all jump naked into a lake shouting 'Bourbaki!'

Computers changed proofs. They were also a catalyst for fascinating new mathematics.

Cell Mates

On a chilly London day, I went to talk to a man about a spaceship. Paul Chapman was sitting outside an Italian restaurant in a dark coat, his Panama hat glowing orange underneath a heat lamp. Dark eyebrows framed large rimless glasses above a scruffy grey beard. Paul belongs to an exclusive group devoted to a mathematical recreation called the Game of Life. He was impatient to tell me its latest discovery.

'Headline news,' he asserted, as he took a black notebook from his pocket and unfolded a tattered piece of paper. 'I carry this around with me wherever I go.' The Game of Life was invented forty years ago by a young lecturer at Cambridge, John Conway, who created the laws of a fantasy universe in which patterns on a square grid evolve and mutate in mesmerizing and unpredictable ways. Objects known as 'wicks', 'guns', 'puffers' and 'spaceships' now populate this universe, and on Paul's piece of paper was the image of Gemini, a spaceship composed of almost a million tiny squares, one of the largest and most sophisticated Game of Life patterns ever constructed. Gemini looked like a diamond made out of herringbone. Paul pointed eagerly at different sections to explain why it was so special. Gemini is the first structure that can self-replicate – it can construct an identical copy of itself. The spaceship is *alive*. Life has finally bred life. 'It's amazing,' he said. 'In forty years we have never seen this before.'

The idea that a mathematical grid can produce a pattern worthy of contemplation dates at least to the 'sieve of Eratosthenes', named after the Greek polymath who we met earlier making the first decent estimate of the size of the Earth. His sieve is a machine for finding prime numbers. We count upwards from 1, and on reaching the next available number we eliminate its multiples. (The method is very similar to the process performed by the autistic savant Jerry Newport in Chapter One.) Our first prime is 2, at which point we eliminate all remaining even numbers. Our second prime is 3, so we eliminate all other multiples of three. Four has been eliminated because it is even, which makes the next prime 5, and so on.

The sieve of Eratosthenes can be drawn for 1 to 100 elegantly in a six-row grid, illustrated below. Horizontal lines after 2, and through 4 and 6, strike out the evens, and a line after 3 captures the odd multiples of 3. Two sets of diagonal lines capture the multiples of 5 and 7. No other lines are needed, since when you are sieving for primes up to an arbitrary number n, you only need to look at the multiples of prime numbers up to $\sqrt{n}$. In this case $n = 100$, so we can stop once we get to 10.

1	7	13	19	25	31	37	43	49	55	61	67	73	79	85	91	97
2	8	14	20	26	32	38	44	50	56	62	68	74	80	86	92	98
3	9	15	21	27	33	39	45	51	57	63	69	75	81	87	93	99
4	10	16	22	28	34	40	46	52	58	64	70	76	82	88	94	100
5	11	17	23	29	35	41	47	53	59	65	71	77	83	89	95	
6	12	18	24	30	36	42	48	54	60	66	72	78	84	90	96	

The sieve of Eratosthenes.

The grid is slick and illuminating, since it tells you instantly that all prime numbers must be in the first or fifth rows, meaning that all prime numbers are either one more or one less than a multiple of 6. What is most important to notice, however, and is the reason why we have a sieve in the first place, is that the prime numbers do not appear in any predictable order. If we continued the grid to infinity they would appear to be randomly sprinkled along the first and fifth rows. That the primes are so easily defined, yet their distribution so capricious, is one of the earliest, and deepest, surprises in maths.

In 1963 Stanislaw Ulam, aged 54, zoned out in a lecture and began to doodle on a sheet of paper. He drew a grid of horizontal and vertical lines and numbered the intersections – as you do – starting with 1 in the middle and spiralling out. He must have been really bored, because he then started to circle the primes. Everyone knows the primes have no obvious pattern, so what could he expect to see? Yet Ulam noticed something new and surprising. The primes had a tendency to sit on diagonal lines, as illustrated overleaf, in a pattern now called an Ulam spiral. When he programmed a computer to continue the spiral from 1 to 65,000, the diagonal lines were still there, as were some shadows of horizontal and vertical order. The Ulam spiral offers a tantalizing suggestion of music beneath the random noise.

Ulam was one of the Polish mathematicians who in the 1930s contributed to *The Scottish Book* in Lwów. In 1935 John von Neumann, a Hungarian mathematician at the Institute for Advanced Study in Princeton, invited him to the US, where he moved for good in 1939. Four years later von Neumann made Ulam, then at the University of Wisconsin, a more curious proposal: to come and join him as part of an unidentified project in New Mexico. Ulam took out a New Mexico guidebook from the university library and saw that previous borrowers were colleagues who had disappeared without explanation. After finding out what they were experts in, he realized what he was being asked to do.

And so Ulam joined the Manhattan Project at Los Alamos, where he was pivotal in the development of thermonuclear weapons, and where he also devised new mathematics. He realized that when the behaviour of a physical system was too complicated to predict, he could get a good sense of the outcome by getting a computer to make lots of random guesses, and refining the results using statistics. During a car ride, Ulam explained the technique to von Neumann and the phrase 'Monte Carlo method' was coined. To find the odds of, say, a roulette ball landing on black, a gambler doesn't need to solve any equations, he can just tally the number of times the ball lands on black after hundreds of random plays. Monte Carlo methods are now crucial in many areas of science. In his downtime

100	99	98	97	96	95	94	93	92	91
65	64	63	62	61	60	59	58	57	90
66	37	36	35	34	33	32	31	56	89
67	38	17	16	15	14	13	30	55	88
68	39	18	5	4	3	12	29	54	87
69	40	19	6	1	2	11	28	53	86
70	41	20	7	8	9	10	27	52	85
71	42	21	22	23	24	25	26	51	84
72	43	44	45	46	47	48	49	50	83
73	74	75	76	77	78	79	80	81	82

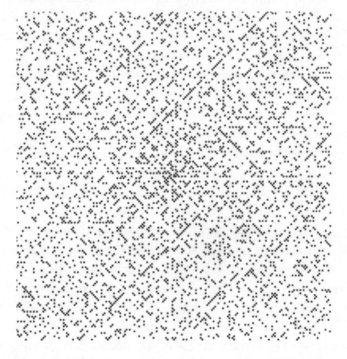

Ulam spirals, from 1 to 100 (top), and from 1 to 65,000 (bottom).

at Los Alamos, Ulam relaxed by inventing solitaire games based on patterns on a grid. By changing the rules that generated the patterns, the shapes could be made to grow and change in interesting ways.

Ulam and von Neumann were best friends, Eastern European émigrés with upper-middle-class, Jewish roots, united by political circumstance and formidable intellects. Von Neumann, it is often said, is the mathematician who most shaped the modern world: he is one of the fathers of the computer, the nuclear bomb, and game theory, the maths of decision-making. His personality matched his achievements. At Princeton he was famous for throwing the biggest parties, during which he often slipped away to his study, because he liked working to the sound of parties.

Von Neumann was both fascinated and scared by the potential consequences of the machines he was building. In the years after the Second World War, science fiction novels and Hollywood movies portrayed a future in which robots took over the world. Von Neumann wanted to know what it would take for a machine to reproduce. He performed a thought experiment involving a robot floating on a lake with an eye and a mechanical arm that could pick up components and build a new version of itself. But the experiment got bogged down with mechanical complications. Ulam suggested that to focus purely on the logical aspects of replication he should not consider anything real like a machine, but instead look at patterns on a grid, just like the solitaire patterns he had played around with at Los Alamos. As a result of the conversation the two men came up with a new mathematical concept, the 'cellular automaton'. Essentially this is a checkerboard grid containing cells, in which the behaviour of each cell depends only on the state of its neighbours. Von Neumann designed a cellular automaton in which each cell was in one of 29 possible states, and came up with the theory for a pattern of 200,000 cells that would self-replicate. Cellular automata became objects of minor academic interest until a decade later, when they caught the eye of a British mathematician with an even more playful mind than Ulam's.

During the 1960s, the maths faculty common room at Cambridge University resembled an after-school club. Staff and students were always playing board games, and devising them. There were so many ideas that one don even kept a file, Games Without Names, together with its companion, Names Without Games. Thriving in this milieu was John Conway, Liverpudlian backgammon fiend and rising mathematical star. One of Conway's inventions was a cellular automaton on a square grid he called the Game of Life. The word 'game', however, was misleading. There were no winners, losers, or even players. The Game of Life was a two-dimensional universe governed by four laws. The point of the game was to set up a starting configuration, the initial pattern, and observe it evolve.

In Life, a cell is either alive or dead, and obeys the following rules:

Birth: a dead cell with *exactly three* live neighbours becomes alive.

Survival: a live cell that has *two or three* live neighbours stays alive.

Death by loneliness: a live cell with *zero or one* live neighbour(s) dies.

Death by overcrowding: a live cell with *four or more* live neighbours dies.

The small print: each cell has eight neighbours, which are the four cells adjacent to it and the four it touches diagonally at its corners. The laws are applied simultaneously to all cells, and each application signifies a new generation.

That's it. There is nothing more to the Game of Life.

Conway decided on birth, death and survival rules so that patterns neither died too quickly nor grew too fast, but instead behaved as interestingly as possible. Imagine one live cell on its own. It dies from loneliness after one generation. Likewise, a pattern consisting of two neighbouring cells also dies after one tick of the clock. But when we start to consider shapes made from three live cells, the organisms are resilient enough to survive, at least for a bit. The illustration opposite shows what happens to the chevron made of three live cells. (Live cells are black, dead cells white.) The two live cells on the base each have only one live neighbour, so they die when we apply the laws. The live cell on top has two live neighbours, so it survives, and the dead cell in the middle has three live neighbours so it becomes alive. After one generation, therefore, the chevron becomes a column of two live cells, and after another it dies.

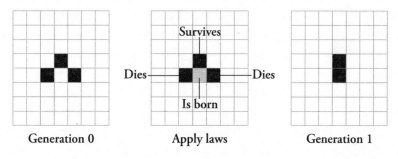

| Generation 0 | Apply laws | Generation 1 |

How the chevron evolves.

The fate of four other triplets is illustrated below. (Each new generation is drawn underneath the previous one. In reality, of course, each new generation occupies the same cells.) By the second generation, two have expired. The square of four cells, however, which Conway called the 'block', lives, remaining unchanged in all

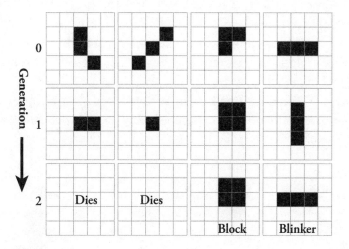

The fates of the triplets.

subsequent generations. The line of three live cells alternates between horizontal and vertical, and is known as the 'blinker'. Patterns that do not change, like the block, or which oscillate between fixed patterns, like the blinker, are called 'stable forms'.

We glimpse the magic to come when we consider the Life histories of the five 'tetrominoes', the cast of the computer game Tetris, which are made from four live cells all connected adjacently. The block, as we saw previously, is inert. The four others are illustrated opposite. After two generations the 'I' and the 'S' become the stable form called the 'beehive', which the 'L' becomes after three. On the other hand, the 'T' has an explosion of energy, evolving through nine generations to a final pattern of four blinkers, 'traffic lights'.

The fun in the Game of Life was its unpredictability. There was no way of knowing what would happen even to simple shapes without tracking them through the generations, which Conway and his colleagues did by hand. Live cells were counters placed on the board of the Oriental game Go, a 19 × 19 square grid. When the pattern required more space, additional boards were put on the floor. New stable forms were found, which received names like 'loaf', 'ship', 'boat' and 'snake'. Sometimes an initial pattern died off, or changed quickly into one of the familiar stable forms. And sometimes, to everyone's excitement, it erupted into life. The R-pentomino, ![R-pentomino], for example, was made of only five live cells, but kept on evolving for dozens of generations until, at the 69th generation, a remarkable event took place. The pattern gave birth to a five-cell pattern that appeared to glide across the board.

The new shape became known as the 'glider', and its behaviour is illustrated below. After two generations it flips over, and after

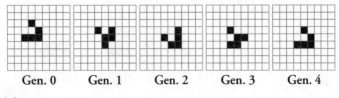

| Gen. 0 | Gen. 1 | Gen. 2 | Gen. 3 | Gen. 4 |

The glider.

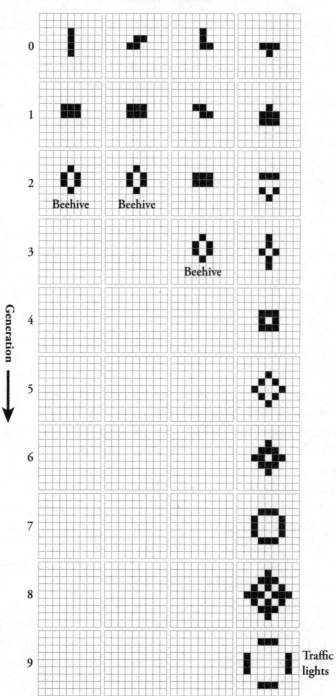

The fates of the tetrominoes.

another two it flips back in such a way that it is one cell down and one cell across from its initial position. The glider will keep shifting one down and one along every four generations. It will continue moving in the same diagonal direction, ad infinitum, so long as nothing is in its path. Conway, the taxonomist-in-chief, designated a new Life species to patterns like the glider that move in straight lines. He called them 'spaceships'.

In 1970 the journalist Martin Gardner wrote about the Game of Life in his longstanding *Scientific American* column, helping to turn Conway's mathematical recreation into one of the earliest international computer crazes. Tracking Life histories using a Go board was time-consuming and error-prone. With computers, the patterns could not only be monitored for much longer, but as the screen flashed through the generations they looked like they were alive. A grid scattered with live cells provided a primordial broth of volatile, transmuting configurations. The R-pentomino, for example, fizzed and frothed for an incredible 1103 generations, leaving the detritus of a ship, a boat, a loaf, four beehives, four blinkers, six gliders and eight blocks. Life was easy to program since there were only four rules, yet it produced complex behaviour not seen on computers before. Designing patterns and seeing how they fared was such an addictive pursuit that during the seventies Gardner estimated it cost US industry millions of dollars in lost computer time. One reader told Gardner that he installed a secret switch under his office desk to flip his computer screen to Life when colleagues left the room.

At MIT, the Game of Life became a *way* of life. A tight fraternity of anarchic, jokey brainiacs made it their mission to explore the toy universe more deeply than anyone else. These were the first computer hackers, the original techno-geeks. (The word 'hack' had first been used in MIT's model railway club to describe an alteration implemented for pleasure alone, but it then came to mean any piece of wildcat programming. Only years later did it gain its current meaning of digital trespass.) The communitarian, anti-authoritarian attitude of the hackers influenced America's emerging computer culture, setting the tone for later innovators such as Steve Jobs and Bill Gates. 'The plan was simply to collect wildlife and domesticate it,' said Bill Gosper, king of the hackers, now a maths tutor in Los Altos, California. Gosper would spend all night playing with Life in the MIT computer room, and it became his routine for almost two years.

Conway set a challenge in the pages of *Scientific American*, for which he offered a $50 prize. Was there a pattern that kept on growing, whose total number of live cells increased without limit? Gosper found one, and pocketed the cash. The 'glider gun' is a pattern of 36 live cells that pulsates in and out, like a beating heart, giving birth to a glider every 30 generations. The gliders scamper away in a diagonal line, one after another, like an endless stream of bullets shot from a gun. Discovery of the glider gun changed the focus of Life study from zoology to physics. No longer were Gosper and his entourage natural historians, passively inspecting the flora and fauna. The fun was in ballistics, devising patterns that contained glider guns firing at other shapes. It was possible to fire two gliders into each other in such a way that they were both annihilated, leaving no debris, as if they had magically vanished into the ether. 'We were trying to find out how to construct things by smashing gliders at each other in every possible way and seeing what happened,' said Gosper. 'And then, what are all those things that you can make by smashing gliders into things you make when smashing gliders?' By doing this Gosper discovered a new seven-cell stable form called the 'eater'. When a glider flies into an eater, the glider disappears in the collision, but the eater reconfigures itself into what it was before, giving the impression that it has ingested the glider. The eater also devours other stable forms positioned near it, always self-repairing after the initial interaction.

The eater was the first indication that Life might have a real-world application, such as in the design of objects that can repair themselves. Not that Gosper was at all interested in that. For him, the glider gun and the eater enabled Life to enter a new phase of engineering *grands projets* in which gigantic patterns could be made up of hundreds of gliders bouncing around between components, with eaters strategically placed to hoover up unwanted debris. One of his early triumphs was the 'breeder', a pattern that breeds gliders. It starts with about 50 of them, and accelerates production so fast that at around generation 6500 the number of gliders exceeds the number of generations.

As the knowledge pool grew, Life enthusiasts engineered ever more incredible patterns. One of my favourites is a simulation of the sieve of Eratosthenes, the ancient mechanism for identifying prime numbers. The Life sieve is made predominantly from guns, gliders and eaters. Its initial configuration contains 5169 live cells. The main component is a gun that fires a 9-cell pattern called the 'lightweight spaceship' horizontally at regular intervals. The spaceships are bombarded by gliders, and the only ones that survive are the second, the third, the fifth, the seventh, the eleventh, and so on – in other words, those whose positions are prime numbers. (A detailed explanation of how it works, with illustrations, is in Appendix Eight on p. 301.)

I love the Life sieve because it transforms the most ancient mathematical algorithm into an intergalactic shootout between flotillas of gliders and spaceships. It's like watching an epic sci-fi battle scene, or, perhaps, the evolution of a colony of particularly mathematical ants. Remember, once you have constructed the original pattern, you never interfere with it again. The pattern could theoretically continue for ever, firing out spaceships in the positions of the prime numbers ad infinitum.

'The ingenuity is staggering,' said Gosper of the best patterns. 'People who try [to make patterns] quickly realize how difficult it is and when they see a success they admire it enormously. You need to be almost insanely manic in order to concentrate hard enough.' Since the seventies hundreds of other dazzling patterns have been created, including one that will calculate the value of pi, invented by a British teenager, Adam P. Goucher. What is there left to achieve? 'Life is an inexhaustible supply of questions and problems,' Gosper replied.

The Game of Life challenges our preconceptions about how the world works because it shows how a simple set of local rules can generate incredibly complex overall behaviour. When you see such a beautifully integrated system as the sieve of Eratosthenes, it is stunning to think that each cell is paying attention to only its eight neighbouring cells.

The Game of Life also demonstrates how different scales present different realities. The sieve of Eratosthenes is a piece of engineering based on the physics of glider guns. It uses the technology of collisions and spaceships. Yet at a granular level there is no such thing as a 'collision' or a 'spaceship'. There are only stationary squares that are either 'alive' or 'dead'.

As more complicated patterns were devised, the question became: What is the limit of what a Life pattern can do? Remarkably, it can do anything your PC, tablet or phone can. If a task can be carried out by a computer, then it can also be carried out by a pattern in Life.

Conway proved this statement by showing that it is possible to build a 'Life computer', meaning an initial configuration of live cells that emulates a computer's internal circuitry. You'll have to take my word for this (or read a book on computing), but computer circuitry at its most basic level is made up of the following components: wire, logic gates and a memory register. A clock sends electronic pulses around the circuit, and these represent binary numbers. A pulse is a 1 and the absence of a pulse is a 0. Conway had the insight that gliders could represent pulses travelling down a wire. The presence of a glider is a 1, and the absence of a glider is a 0. A stream of gliders can therefore represent any number consisting of 0s and 1s, as illustrated below. Since gliders move diagonally, I have moved the grid to an angle of 45 degrees.

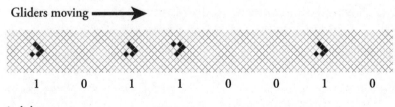

Gliders moving

| 1 | 0 | 1 | 1 | 0 | 0 | 1 | 0 |

A glider stream.

Conway constructed the simplest type of logic gate, called a NOT gate. A logic gate is a component that has an input and an output.

Some wires go in, some wires come out. The NOT gate has only one input wire and only one output wire. The output contains the reverse signal of the input: it changes 1 to 0, and 0 to 1. A NOT gate in Life, therefore, must turn the presence of a glider in the input stream into the absence of a glider in the output stream, and vice versa. Conway saw that a strategically placed glider gun would perform this function, as shown opposite. The input stream is moving horizontally from left to right. The glider gun is firing gliders vertically downwards. If there is a glider in the input stream, it will be annihilated by a glider coming from the gun. But if there is no glider in the input stream, a glider from the gun will pass through unscathed because there is nothing for it to collide into. The output stream therefore contains a 1 if there is a 0 in the input, and a 0 if there is a 1. It's a NOT gate. The output is at right angles to the input, but this doesn't matter since we can change the direction of the stream later if need be.

All logic gates are combinations of three basic types: NOT, AND and OR. Conway constructed patterns made from guns and eaters that emulated AND and OR gates too. He showed that glider streams could be made to change direction, thus emulating the bending of wires. He showed how glider streams could be thinned out, so two streams could cross each other without fear of collision, thus emulating the crossing of wires. And he showed how a memory register could be made up of blocks. Each block stores a number, depending on its distance from a certain point. Gliders crashing into the block move it nearer or further away from that point, changing its value. This completed Conway's proof: by being able to construct wires, logic gates and a memory register in Life, he had shown that it was theoretically possible – given a large enough grid – for his mathematical pastime to emulate any computer in the world.

John Conway lost interest in Life after he worked out the above proof. (In 1986 he moved to Princeton to take the John von Neumann chair of mathematics.) But many enthusiasts had an addiction for which there was no cure, and which would only deepen. The international community of Lifers numbers around

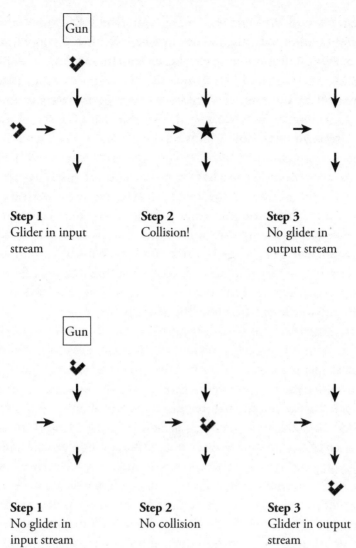

A NOT gate is made up of a gun firing gliders perpendicular to the input stream.

a hundred, and includes Paul Chapman, who at the turn of the century decided it was time to build the Life computer. 'There is a difference between knowing something can be done, and doing it,' he said.

Like many who share his passion, Paul is not an academic. He read maths at Cambridge in the seventies – attending Conway's lectures – and then became an IT consultant. Paul now lives in central London, near the restaurant where we met. He had insisted on meeting outdoors, despite the inclement weather, because he objects to the ban on indoor smoking. He rolled his own cigarettes as we spoke. 'The reason I like Life is that you are always astounded by it,' he said. 'Whenever you look for better ways of doing something you find dozens of ways.'

Just as a computer has hardware and software, a Life pattern emulating a computer also has 'hardware' and 'software'. The former emulates the wiring of the machine and the latter the program that the machine will read. For his Life computer, Paul did not use Conway's circuitry of guns, gliders and eaters but a newer, more efficient technology based around a seven-cell pattern called the Herschel. His pattern contained several million live cells and a program instructing it to calculate 1 + 2. 'Adding 2 + 3 would have taken far too long,' he said. The pattern began with a spaceship hitting a stable form, which triggered a signal that was sent off to collide with various components, which themselves triggered more signals, and as the signals worked their way through the machinery the whole pattern resembled a giant game of bagatelle. Eventually, a block on the output register displayed the number 3. 'I was so excited,' he said. 'If I can add one and two, then I know that this same machine can calculate the millionth digit of pi, run Windows, or, if you gave it the right parameters, emulate the life cycle of a star!'

Of course, Paul's Life computer was of no practical use for doing any of these things. But it returned Life to its roots. John von Neumann came up with the idea of cellular automata to investigate self-replication. Paul's pattern opened up the tantalizing possibility of self-replicating creatures in Life.

Superficially, patterns evolving on a Life grid look alive because they morph and twist organically as you move through the generations. For something properly to be alive, however, it needs to be able to replicate itself. But what is replication? The glider replicates, for example, in a trivial way. It is a five-cell pattern that has an identical shape every fourth generation, moved one square down and one across. What von Neumann wondered was how a computer might *build* an identical copy of itself. To do this he needed to solve a mathematical puzzle, since there is a logical paradox contained within the mechanics of self-replication.

We saw above that computers consist of hardware and software. Let's call the hardware the 'constructor'. And let's call the program that we feed into the constructor, which tells it to build a copy of itself, the 'blueprint'. Our hope is that when we feed the blueprint into the constructor, it builds a new constructor together with a new blueprint, thus replicating the two original elements. But there is a problem: *does the blueprint contain instructions on how to build a new blueprint?* If it does, then the instructions must also contain instructions on how to build a new blueprint, which must in turn contain instructions on how to build instructions on how to build a new blueprint, and so on for ever. You get an infinite regress of instructions within the blueprint, which is not allowed since the blueprint must be finite. On the other hand, if the blueprint does not contain any information about itself, then the machine is not fully self-replicating, since the new machine will have no blueprint. Before von Neumann could think about engineering, he needed to fix the maths.

Von Neumann realized that for a machine to self-replicate he needed to introduce a new component that duplicates the blueprint, the 'blueprint copier'. So, when the constructor reads the blueprint it builds a new machine that is perfect in every way except one – it lacks a blueprint. The final stage is for the copier to duplicate the blueprint and send it to the new machine. Von Neumann's self-replicating machine therefore uses the blueprint in two different ways: the constructor *reads* it like a set of instructions, and the copier *duplicates* it. Only by treating the blueprint once as a code and once as an object was it possible to eliminate the infernal riddle of infinite regress.

The pattern von Neumann designed for his original cellular automaton contained a constructor, a copier and a blueprint. He showed in theory that it would self-replicate, but did not demonstrate this practically because computers were not powerful enough. Nevertheless, his work influenced a generation of computer scientists, philosophers and even biologists, who in the 1950s were studying how living cells reproduce. When they did work it out, over that decade and the next, they discovered that von Neumann had been right! He had correctly anticipated the general framework of organic reproduction. Every cell has a blueprint, its DNA, which contains coded instructions for creating new cells. But DNA does not contain a description of itself – the DNA that emerges in a new cell is the result of replication. (The double helix splits in two, and enzymes recreate two identical copies of the original.) Just as von Neumann's blueprint is read in two different ways by a machine, so DNA behaves in two different ways when a living cell reproduces.

Paul Chapman tried to design a self-replicating pattern, but he couldn't figure out how to duplicate the blueprint. In 2010, however, the Canadian programmer Andrew Wade unveiled Gemini. 'When I first saw it, it blew me away!' said Paul. 'Gemini is the most important pattern for forty years. And no one knew who Andrew Wade was! He just announced it on a bulletin board!'

Gemini is the first self-replicating pattern in the Game of Life. It is shaped like an enormously long and thin dumbbell, as sketched opposite. At each end are identical constructors (hence the pattern's name), and between them – stretching across a 4 million × 4 million grid of cells – is the blueprint, made up of gliders. Wade's ingenious idea was *not* to duplicate the blueprint, but instead to have it snake between the constructors. When the blueprint reaches one of the constructors, it tells the constructor to build a new version of itself 5120 cells up and 1024 cells along, and to destroy itself at the same time. The blueprint is then sent back in the opposite direction where, a few million cells later, it reaches the opposite constructor, telling it also to build a new version of itself 5120 cells up and 1024 cells along, and then to autodestruct. The cycle repeats every 33.7 million generations, at which point the entire pattern has moved into a different position, 5120 up and 1024 along. Because Gemini moves, the pattern is considered a spaceship, but it does not move by flip-flopping along like the glider; it moves through a process of self-replication. 'The brilliant thing that Andrew Wade did,' said Paul, 'was to eliminate the [blueprint] copy step, and just have

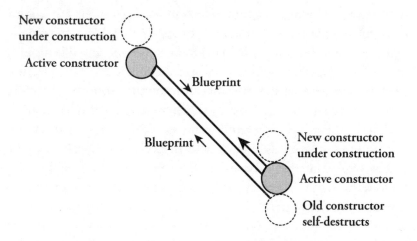

New constructor under construction

Active constructor

Blueprint

Blueprint

New constructor under construction

Active constructor

Old constructor self-destructs

Gemini is the first spaceship whose motion is based on self-replication.

[the blueprint] turn up, as it were, out of the blue, at just the right moment to give its instructions.'

Gemini also represents another important breakthrough: it is the first spaceship that moves obliquely, meaning neither horizontally nor vertically, nor at 45 degrees to the grid.

Paul showed me a piece of paper with one of Gemini's constructors. He mentioned, proudly, that it was based on his Life computer. The image looked like a geometric smudge, ordered grey chevrons surrounded by tiny dots. I asked if he had a picture of Gemini in its entirety. He replied that there would be no point, since at that scale it would be so thin as to be invisible. Almost all of the pattern is the glider stream. Perhaps counter-intuitively, the blueprint takes up much, much more room than the constructor. Von Neumann realized this imbalance too: the constructor in his automaton fits into a 97 × 170 array, but the blueprint is 145,315 cells long. Large patterns consist mostly of empty space. 'Maybe the reason there is so much space in the Game of Life is for the same reason that there is so much space in our world,' said Paul. 'Atoms have to have room to do their thing.'

Gemini has heightened anticipation about the next stage of Life exploration, especially the prospect of a pattern that self-replicates with variation. If a pattern produces copies that have minor differences, Darwinian natural selection will follow. In 1982 John Conway speculated that on a large enough Life grid in an initially random state it was probable that 'after a long time, intelligent, self-replicating animals will emerge'. Three decades later, his conjecture still quickens the pulse of the Life community. Some of the most exciting work is being done by Nick Gotts, a complex-systems analyst in Aberdeen, Scotland, who is finding new patterns by randomly populating Life grids with live cells. He calls his project 'sparse Life', since the proportion of live cells to dead ones has to be low, otherwise there are too many uncontrolled interactions. 'In some of the patterns I have worked on it looks like there is something akin to natural selection,' he said. 'There are patterns that modulate the emergence of other similar types of pattern. I am convinced that, if you let my program run long enough, natural selection would emerge.'

Cellular automata more basic than the Game of Life can produce behaviour that is equally complex. Let's consider an automaton in *one* dimension: a row of cells in which each cell has only two neighbours. Each cell can be either alive or dead.

Consider the following rule:

If both of a cell's neighbours have the same state then the cell is dead in the subsequent generation. Otherwise it is alive in the subsequent generation.

The rule is illustrated below. The image shows the eight possible combinations of a cell and its two neighbours. Underneath each combination is the state of the cell after one generation. The first combination has a live cell surrounded by two live neighbours. It is dead after one tick. The second combination has a live cell on the

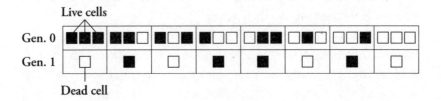

left and a dead cell on the right. The middle cell survives after one tick. And so on. When the two neighbours are the same colour, the cell below is white. When the two neighbours are of differing colours, the cell below is black.

One way to think about this rule is to consider a group of people who stand in line for the bus every morning in the same order. Each person has two neighbours, one at either side. Let the rule concern the wearing of hats: if both your neighbours are wearing a hat, then hats are too common, and you will not wear a hat the following day. If neither neighbour is wearing a hat, then hats are out of style, and neither will you wear a hat the following day. If, however, only one neighbour is wearing a hat, hats are neither tacky nor passé, and the next day you will wear a hat. The cellular automaton provides a model for daily fluctuations in headgear.

To illustrate the behaviour of one-dimensional cellular automata we draw a row with a single live cell (generation 0), and then apply the rule to every cell to generate a new row underneath (generation 1). We then apply the rule on every cell in this row to get a new row (generation 2), and so on. The image below shows what happens. (Note that the apex of the triangle is the live cell on the first row, and each row below it is a new generation, unlike the Game of Life, in which the whole grid is the same generation. I have omitted the grid, so the pattern is seen more clearly.) The result is the beautiful mathematical ziggurat known as the Sierpinski triangle, a fractal structure of nested triangles.

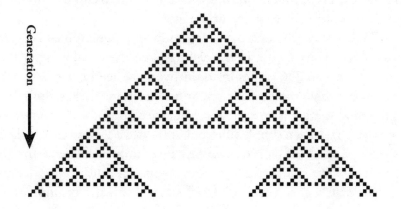

277

Because there are 8 combinations of a cell and its two neighbours, and 2 possible states (alive or dead), there are $2^8 = 256$ different sets of 'genetic rules' for one-dimensional cellular automata. They are numbered from 1 to 256. The one on the previous page, known as Rule 90, produces orderly patterns. Others, like Rule 30, are weirder. This rule and the pattern it produces starting from a single live cell are illustrated opposite. The shape contains a mixture of regular and organic-looking areas. The serrated crust on the left slope displays order. As we move to the right, however, we see a disorderly, pockmarked rock-face of oddly sized triangles.

Stephen Wolfram has the Rule 30 pattern embossed on his business cards. When I met him, he took one out of his wallet and handed it to me. We were sitting in the headquarters of his company, Wolfram Research, in Champaign, Illinois. Wolfram has the face you might expect of a child maths prodigy now in middle age: round and pale, with tufts of hair around a professorial crown. When he spoke he often gazed into the middle distance, his eyes, behind glasses, oscillating rapidly like an electronic display indicating that his brain was operational. Wolfram started young, publishing his first scientific paper while a schoolboy at Eton in the 1970s. By his early twenties he was at the Institute of Advanced Study in Princeton. An early convert to computers, he devised a computer language which became the basis of *Mathematica*, software for drawing curves and solving equations, which is now used throughout education and industry. Since 1987 he has run Wolfram Research, which, thanks to the success of *Mathematica*, has enabled him to conduct his own research independent of any university.

Wolfram was the first person to study one-dimensional cellular automata in any depth, in the eighties, and the numbering of the rules from 1 to 256 comes from this work. When he saw Rule 30, it was a thunderbolt to his scientific intuition. '[It] is the most surprising thing I have ever seen in science,' he said. Wolfram was amazed that such a simple rule could produce a pattern that looked so complicated. He looked closely at the column directly underneath the initial live cell in the first row. Assuming a live cell is 1 and a dead cell 0, the sequence of cell states down the column went

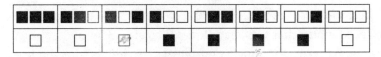

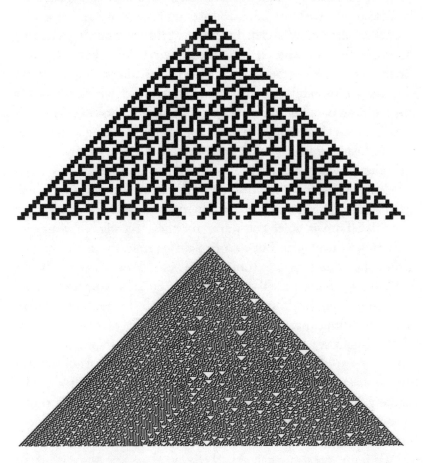

Rule 30: its genetic laws, its evolution after 50 generations, and its evolution after more than 200 generations.

1, 1, 0, 1, 1, 1, 0, 0, 1, 1, 0, 0, 0, 1, 0 … There was no pattern to it. To his astonishment, the sequence was shown by standard statistical tests to be perfectly random. Rule 30 is defined deterministically, yet the pattern in its central column is so unpredictable that it is indistinguishable from a succession of coin flips. (Wolfram patented Rule 30 as a random-number generator, and it is used in *Mathematica*.)

Wolfram was also captivated by another rule, Rule 110. It produced a grid that was again a mixture of regular and apparently random patterns – enough complexity, he conjectured, for it to be able to emulate the circuitry of a computer, in the same way that the Game of Life can. In 2004, Matthew Cook proved that his conjecture was true. It is theoretically possible, therefore, for a single row of cells to be able to perform anything a computer can do, using only one set of rules that determines whether a cell is alive or dead based on the state of its two neighbours. Likewise, it is possible for a single row of people to perform anything a computer can do, using only one set of rules that determines whether or not they should wear hats.

Cellular automata are discrete mathematical models in which fixed local rules generate surprisingly complex behaviour on a larger scale. Wolfram is chief flag-waver for the view that they are not only fun to play with but also explain complexity in the natural world. His opinions are 'summarized' in *A New Kind of Science*, a 1280-page book he self-published in 2002. In it he argues that the insight you get from looking at Rule 30, for example, reveals a new scientific paradigm. Consider the cone of the venomous cloth-of-gold sea snail, illustrated opposite. The standard view of evolution explains the pattern as the result of natural selection. But just look at the picture of Rule 30! 'This is the thing I consider striking,' said Wolfram. 'You just have to pick these simple [cellular automata rules] at random and you'll get something like [the pattern on the shell].'

Wolfram goes further. He believes that the universe at a fundamental level is a cellular automaton. In other words he thinks that the structure of the universe is analogous to a grid in the Game of Life, but one that exists beyond space and time. So you reading this book is the nth generation of an initial configuration of cells that has evolved under a small set of local rules. Wolfram has set himself a mission to discover these rules. 'If it turns out that they are the equivalent of three lines of code and we didn't look for those lines in this century, it's kind of embarrassing,' he said.

Wolfram is not the only academic to believe that the universe may be a cellular automaton, but he is the one who has spent most

The shell of a cloth-of-gold sea snail, or Conus textile.

time and money trying to prove it. He has been systematically road-testing sets of rules to see the types of universe they produce. 'For a while I had this very quaint thing that I could say that I had this computer in my basement searching for the universe.'

He described his strategy: 'You look through these different, very, very simple sets of rules and some of them are obviously hopeless. Like the universe dies out after two steps. Or the universe expands forever in a way where no piece of the universe ever communicates with any other piece of the universe. All kinds of pathologies. And so you just keep knocking these universes out, so by the time you're at the thousandth universe you're starting to find ones that you can't knock out easily.' He added that he had found universes that were 'not obviously *not* our universe' but that due to the distractions and challenges of running a tech company he had been side-tracked by other projects. Still, he intends to return to universe-hunting in the future. 'I hope that one day I'll have a business card that has the rules of the universe on the back.' He laughed. '*That* would be a good business feature.'

Whether or not the universe is a cellular automaton, the concept is increasingly used in science to model a variety of phenomena, such as traffic flow, the spread of algae on a lake, and the growth of cities:

a cell can be a stretch of road, a piece of lake, or a patch of land. Another application, pioneered by Craig Lent at the University of Notre Dame, is the development of 'quantum-dot cellular automata', in which tiny 'quantum dots' change their electrical charge based on the configuration of neighbouring dots. Lent hopes this nanoscale technology will eventually replace the transistor, since a transistor made from dots will be smaller and will dissipate much less heat than one with traditional wiring. If quantum-dot technology is successful, the cellular automaton may one day be omnipresent in all electronic goods.

John von Neumann and Stanislaw Ulam devised the cellular automaton to solve a problem inspired by the real world: what it would take for a computer to build an identical copy of itself. The prospect of a future with self-replicating machines is bone-chilling. John Conway, however, took the idea and turned it into a whimsical and absorbing mathematical recreation. Automata have subsequently been reinvented, and have found uses with no connection to self-replication. The process is a familiar one: mathematicians are nourished by problems in the world, they play around with the concepts for fun, and sometimes – maybe years, maybe centuries or maybe even millennia later – new applications are found. Only with fresh mathematical insights can technology progress, and science become more able to explain the world around us. At the beginning of this book I said that maths is a joke. I'd like to rephrase that comment. Maths is, and always has been, a game.

It's *the* game of life.

Glossary

Axiom: a statement that is assumed to be true, and from which other statements are deduced.

Benford's law: the phenomenon that in many naturally occurring data sets the chances of a number beginning with a 1 is about 30.1 per cent, the chances of a number beginning with a 2 is about 17.6 per cent, and so on.

Bisector: a line that cuts another line in half.

Calculus: the umbrella term for differentiation and integration, which are the mathematical tools required to analyse quantities that vary with respect to each other.

Cartesian coordinates: a map of the plane in which every point is determined by a horizontal and a vertical position. Usually the Cartesian plane is drawn with two perpendicular axes intersecting at the point (0, 0).

Cellular automaton: a mathematical model consisting of discrete cells that change their state every unit of time depending on the state of neighbouring cells.

Chord: a line between two points on a circle.

Circle constant: another term for pi, the circumference of a circle divided by its diameter.

Common fraction: a fraction written with a numerator and a denominator, like $\frac{1}{2}$ or $\frac{457}{3}$.

Complex number: a number of the form $a + bi$, where a and b are real numbers and i is $\sqrt{-1}$.

Complex plane: a geometric interpretation of the complex numbers, analogous to Cartesian coordinates, in which the horizontal axis represents real numbers and the vertical axis represents imaginary numbers.

Conic section: the curves made from the intersection of a plane with a cone – the circle, the ellipse, the parabola and the hyperbola.

Conjecture: an unproved theorem that is assumed to be true.

Constant: a number that is fixed, typically used in opposition to a variable, which can take on many values. See also *mathematical constant*.

Continuity: the field that deals with mathematical concepts that are continuous, such as lines and areas.

Continuous: generally, a continuous curve is an unbroken one, like a line drawn by a pencil across a page. The strict definition uses the concept of a limit, which is too detailed and advanced to explain here.

Curvature: the measure of how much a curve deviates from a straight line.

Cycloid: the path made by a point on the edge of a wheel rolling along a straight line.

Derivative: a formula representing the gradient of a curve, or equally the rate of change of a variable quantity.

Differential equation: an equation that includes either derivatives or integrals.

Differentiation: the process of transforming a curve into its derivative.

Distributive law: a basic rule of arithmetic, which states that for the numbers a, b and c, then $(a + b) c = ac + bc$.

***e*:** the exponential constant, which begins 2.718.

Eccentricity: a measure of how much a conic section deviates from a circle.

Equilateral triangle: a triangle with three equal sides.

Exponent: see *power.*

Exponential constant: the number beginning 2.718 that has the symbol *e.*

Exponential growth/decay: when the growth/decay rate of a quantity is a fixed proportion of the total amount.

Factorial: the multiplication of a whole number by every whole number less than it. So, for example, the factorial of 5, written 5!, is $5 \times 4 \times 3 \times 2 \times 1 = 120$.

Fast Fourier Transform, or FFT: an algorithm that allows the rapid calculation of Fourier series.

Focus: a significant point used in the geometrical construction of conic sections.

Fourier series: the sum of (a possibly infinite number of) sinusoids which, when added together, make the wave under discussion.

Fourier transform: the process of transforming a periodic wave into its Fourier series, and also the name given to the series.

Fractal: an object that contains elements of self-similarity.

Fundamental Theorem of Algebra: the theorem that all polynomial equations can be solved, and that the solutions are always complex numbers.

Fundamental Theorem of Arithmetic: the theorem that every whole number greater than 1 is either prime or the product of a unique combination of prime numbers.

Fundamental Theorem of Calculus: the theorem that integration is the reverse operation to differentiation, and vice versa.

Gradient: the mathematical measure of slope, or the rate of change of vertical distance with respect to horizontal distance.

Harmonograph: a drawing machine in which a stylus moves with simple harmonic motion in at least two non-parallel directions.

Hypotenuse: the line opposite the right angle in a right-angled triangle.

i: the symbol for $\sqrt{-1}$.

Imaginary number: any multiple of i.

Integral: a formula representing the area underneath a curve, or equally the rate of accumulation of a variable quantity.

Integration: the process of transforming a curve into its integral.

Limit: if a sequence of values gets closer and closer to a fixed value, such that the sequence becomes as close to that value as you want it to be, then the fixed value is the 'limit' of the sequence.

Locus: a curve made up of points that satisfy a given mathematical constraint.

Logarithm: for a mathematical definition of logarithms, and the logarithmic scale, see Appendix One overleaf.

Log-log scales: a graph in which both axes are logarithmic scales.

Mathematical constant: a fixed number that arises naturally in mathematics, such as pi or e.

Nomogram: a diagram that allows you to do calculations by drawing a line and seeing where it crosses a scale.

Number line: a geometrical interpretation of numbers laid out in order along a continuous line that extends to minus infinity on the left and plus infinity on the right, with zero in the middle.

Origin: the point (0, 0) on a coordinate graph.

Periodic wave: a wave that repeats itself every fixed period.

Pi: the numerical value of the circumference of a circle divided by its diameter, which begins 3.14.

Polar coordinates: a map of the plane that locates each point by its distance from a fixed point, the pole, and its angle from a fixed direction.

Polygon: a two-dimensional shape with an outline made up of straight lines.

Polynomial equation: an equation containing constants and variables that only uses the operations of addition, subtraction, multiplication and whole number powers. All the equations we learn at school are polynomials.

Power: when a number n is multiplied by itself a times, we write it n^a, and say that a is the power, or exponent, of n.

Power law: two variables follow a power law if one variable is proportional or inversely proportional to the power of the other.

Prime number: A whole number greater than 1 that is only divisible by itself and 1 (for example, 2, 3, 5, 7, 11, 13, 17 …).

Proof: the chain of reasoning used to establish that a theorem is true.

Real number: any point on the number line, which comprises whole numbers, common fractions and those numbers like pi and e that cannot be written as common fractions.

Right angle: a quarter turn, or 90 degrees.

Roulette: a curve made by a point on a rolling wheel.

Scaling law: an equation in which one variable is size and the other varies with size.

Self-similarity: the property of an object that is exactly the same, or nearly the same, as a smaller part of itself.

Set: a group of things.

Set theory: a branch of mathematics concerned with the properties of sets, and how they provide a basis for arithmetic.

Shape: the external geometry of an object independent of how big it is or where it is.

Similar: used to describe two objects that have the same shape but not the same size.

Simple harmonic motion: an oscillation that is sinusoidal over time.

Sine: the trigonometric ratio that comes from dividing the opposite side by the hypotonuse.

Sine wave: a curve generated by the vertical displacement of a point rotating around a circle.

Sinusoid: a curve with the shape of a sine wave.

Tangent: a straight line that touches a curve at a single point. Also, the trigonometric ratio that is the opposite side divided by the adjacent side.

Theorem: a statement that is not self-evident, but which has been proved by deductive reasoning.

Triangulation: the measurement of distance using the trigonometric ratios.

Trigonometry: the field of mathematics derived from studying the ratios between the sides of a right-angled triangle.

Variable: a quantity that can assume different values.

Vertex: one of the corner points of a triangle, or of any other shape made from straight lines.

Whole number: in this book, taken to be the positive numbers 1, 2, 3 …

Appendix One

A logarithm is defined thus:

If $a = 10^b$

then the logarithm of a is b, which is written:

$\log a = b$

In other words, when a number a is expressed as a power of 10, the logarithm of a is the value of that power. Here are some easy logarithms:

$\log 10 = 1$, since $10 = 10^1$
$\log 100 = 2$, since $100 = 10^2$
$\log 1000 = 3$, since $1000 = 10^3$

And here is a table of logarithms of the numbers between 1 and 10:

$\log 1 = 0$ $\log 6 = 0.778$
$\log 2 = 0.301$ $\log 7 = 0.845$
$\log 3 = 0.477$ $\log 8 = 0.903$
$\log 4 = 0.602$ $\log 9 = 0.954$
$\log 5 = 0.699$ $\log 10 = 1$

If we mark the logs of 1 to 10 on a line positioned according to their values, we get the 'logarithmic scale' between 0 and 1. The logs become more bunched up the further along they are:

I have also marked the distances between the logs. You will recognize them as the first-digit percentages on p. 29. In other words, if I choose a point randomly on the line between 0 and 1, there is a 30.1 per cent chance that the point will be in the interval between log 1 and log 2, there is a 17.6 per cent chance of it landing in the interval between log 2 and log 3, and so on.

Equally, the first interval has length log 2 – log 1, the second interval has length log 3 – log 2, and the dth interval has length log $(d + 1)$ – log d. This means that the probabilities can be expressed more concisely as log $(d + 1)$ – log d for each digit d.

Appendix Two

Here I will show that any equation of the form $y = \frac{k}{x^a}$ will always produce a left-sloping straight line on a double logarithmic scale. And vice versa, that a left-sloping straight line on a double logarithmic scale can always be represented by an equation of the above form. If the axes denote the logs of rank and frequency, then a left-sloping straight line produces the Zipf's law equation frequency $= \frac{k}{\text{rank}^a}$.

We need to assume some familiarity with coordinate geometry, such as the concept of gradient, and also some basic properties of logarithms. We also need to accept that the following statement is true:

(1) On a coordinate graph where x and y denote the horizontal and vertical axes, all straight lines can be written in the form $y = mx + c$, where m is the gradient of the line and c is the point where the line crosses the vertical axis.

So, we start with:
$$y = \frac{k}{x^a}$$

Let's take the log of both sides of the equation:
$$\log y = \log \left(\frac{k}{x^a}\right)$$

The properties of logarithms mean that we can expand this to:
$$\log y = \log k - \log x^a$$

And once again:
$$\log y = \log k - a \log x$$

If we let $\log y = Y$ and $\log x = X$, this equation becomes:
$$Y = -aX + \log k$$

From our assumption (1) above, we know that on a graph where X and Y denote the horizontal and vertical axes, this is a straight line with gradient $-a$ that crosses the vertical axis at $\log k$.

Since $X = \log x$ and $Y = \log y$, the graph must be displaying a double logarithmic scale, and since the gradient is negative, we know that the line must be left-sloping.

Likewise, imagine we have a left-sloping straight line on a graph with a double logarithmic scale. From (1) the line can be written in the form:

$$\log y = -a \log x + c$$

(Since the line is left-sloping, we can specify that the gradient is negative.)

If we let $c = \log k$, we have the equation:

$$\log y = -a \log x + \log k$$

$$\text{or}$$

$$\log y = \log k - a \log x$$

Using the properties of logarithms, we can rearrange this to:

$$\log y = \log k - \log x^a$$

And again to:

$$\log y = \log \left(\frac{k}{x^a}\right)$$

Which means that:

$$y = \frac{k}{x^a}$$

And there we have it.

As a corollary, the equation $y = kx^a$ produces a right-sloping straight line on a double logarithmic scale, and any right-sloping straight line on a double logarithmic scale can be represented by such an equation.

Appendix Three

HEIGHT OF A MOUNTAIN

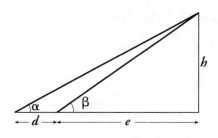

The illustration shows the triangles from p. 70. Our aim is to calculate the height of the mountain, h, with only knowledge of α, β and d. Let e be the distance from the point directly below the summit to the nearest observation point.

We know that $\tan \alpha = \frac{h}{(d+e)}$, and that $\tan \beta = \frac{h}{e}$. Let's rearrange these equations:

$$h = (d + e) \tan \alpha$$
$$h = e \tan \beta$$

So:
$$(d + e) \tan \alpha = e \tan \beta$$

Which rearranges to:
$$e = \frac{d \tan \alpha}{\tan \beta - \tan \alpha}$$

The original equation $\tan \beta = \frac{h}{e}$ rearranges to:
$$h = e \tan \beta$$

Therefore we can say that:
$$h = \frac{d \tan \alpha \tan \beta}{\tan \beta - \tan \alpha}$$

Hence the height is described using only terms with α, β and d.

RADIUS OF THE EARTH

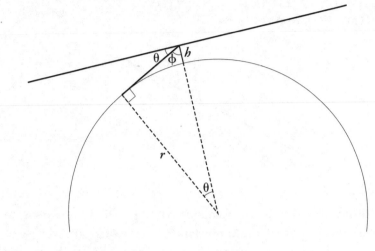

The illustration shows the triangle from p. 71. We know θ, the angle between horizontal and horizon, and h, the height of the mountain. Our aim is to calculate the radius of the Earth, r.

The first step is to show that the angle at the centre of the Earth is θ. We can deduce that the angle ϕ must be $90 - \theta$. Since angles in a triangle must add up to 180 degrees, the angle at the centre must be θ.

We know that $\cos \theta = \dfrac{r}{(r + h)}$

So:

$(r + h) \cos \theta = r$

$r \cos \theta + h \cos \theta = r$

Rearranging:

$r - r \cos \theta = h \cos \theta$

$r\,(1 - \cos \theta) = h \cos \theta$

So:

$$r = \frac{h \cos \theta}{1 - \cos \theta}$$

Appendix Four

THE MULTIPLICATION MACHINE

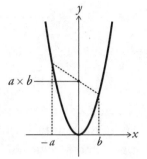

Proposition: In order to multiply $a \times b$, draw a line on the parabola $y = x^2$ from the point where $x = -a$ to the point where $x = b$, as illustrated. The straight line between these points crosses the y-axis at $a \times b$.

Proof: I will have to assume one piece of maths, which is that a straight line through a point with coordinates (p, q) has the equation $y - q = (x - p)m$, where m is the gradient.

The line in the diagram goes between the point with coordinates $(-a, a^2)$ and the point (b, b^2).

The gradient of the line, which is the vertical distance divided by the horizontal distance covered, is $\dfrac{b^2 - a^2}{b + a}$, which can be written $\dfrac{(b + a)(b - a)}{b + a}$, which reduces to $(b - a)$.

So, the equation of the line is:

$$y - a^2 = (x + a)(b - a)$$

We can rearrange this to:

$$y - a^2 = xb - xa + ab - a^2$$

The $-a^2$ terms cancel out, leaving:

$$y = xb - xa + ab$$

When the line crosses the vertical axis, $x = 0$, so:

$$y = ab$$

In other words, the line crosses the axis at ab, which is $a \times b$.

Appendix Five

If a sum S is compounded at rate r, then after t compounding periods, the value of the sum is equal to:

$$S (1 + r)^t$$

The sum will double when $(1 + r)^t = 2$. We solve this equation by taking the natural logarithms of both sides. These are logarithms in base e, and are written ln. So:

$$\ln (1 + r)^t = \ln 2$$

Which reduces to:

$$t \ln (1 + r) = \ln 2$$

So:

$$t = \frac{\ln 2}{\ln (1 + r)}$$

When r is small, $\ln (1 + r) \approx r$, so we can rewrite the equation as:

$$t \approx \frac{\ln 2}{r}$$

Which is the same as:

$$t \approx \frac{0.69}{r}$$

If r is the rate when expressed as a fraction, then let R be the rate as a percentage. We need to multiply top and bottom by 100:

$$t \approx \frac{69}{R}$$

So, the number of compounding periods t it takes for a quantity to double is 69 divided by the percentage growth rate R.

Because 72 is an easier number to divide than 69, the number most commonly used in the rule is 72, even though 69 would be more accurate.

Appendix Six

The largest shaded square has area $\frac{1}{4}$. The second largest shaded square has a quarter the area of the largest square, so has area $\frac{1}{16}$. The third largest square has area a quarter of that, and so on. The total shaded area is:

$$\frac{1}{4} + \frac{1}{4^2} + \frac{1}{4^3} + \dots$$

But for every shaded square, there are exactly two non-shaded squares of equal size. So the shaded squares must also make up a $\frac{1}{3}$ of the total square.

So:

$$\frac{1}{3} = \frac{1}{4} + \frac{1}{4^2} + \frac{1}{4^3} + \dots$$

Appendix Seven

How to divide a cake fairly between three people:

Let's call the three people Hugo, Stefan and Stanislaw, after the three most significant contributors to *The Scottish Book*.

Step 1: Let Hugo make the first cut. He aims to cut out a $\frac{1}{3}$ portion of the cake.

Step 2: Hugo passes this slice to Stefan, who must judge whether, in his opinion, it is $\frac{1}{3}$ or not. If he thinks it is too big, he trims it.

Step 3: The slice is passed to Stanislaw, who can decide to take it or not. If Stanislaw takes it, Hugo and Stefan are left to share the remaining large section of cake plus any of Stefan's trimmings. One of them divides it all into two, and the other chooses.

Step 4: If Stanislaw doesn't take it, there are two possibilities, depending on whether or not Stefan trimmed Hugo's slice.

  (i) If Stefan trimmed the slice, then Stefan must take it. The other two then divide the rest, as in Step 3.

  (ii) If Stefan didn't trim the slice, then Hugo takes it. The other two then divide the rest.

While the method is logically sound, it is potentially messy.

Appendix Eight

Illustration (i) overleaf shows the sieve of Eratosthenes at generation 0. Illustration (ii) shows it at generation 650, when the (prime) numbers 2, 3, 5, 7 and 11 have reached safety, and illustration (iii) is a more detailed look at the corridor of glider fire in (ii).

The procedure is as follows: the pattern highlighted in (ii) marked Gun A fires spaceships that move from left to right, each one representing an odd number. They will leave the main structure if they manage to avoid being shot at by the guns lined up immediately above them.

Let's look more closely at these guns. Going from right to left, which is the order they are created, the first, Gun B, fires a glider diagonally down and left every three intervals. This gun will annihilate all spaceships representing numbers divisible by three. The second, Gun C, fires a glider every five intervals, annihilating all spaceships representing numbers divisible by five. The next gun will annihilate all spaceships representing numbers divisible by seven, and so on. To generalize: when Gun A produces a spaceship representing the odd number n, the corridor expands leftwards in order to create space for a gun that fires gliders at an interval n. The combined effect is that only spaceships in prime number positions will get through. If n is not prime, then it has at least two divisors and the spaceship in position n will eventually be annihilated by the gun firing gliders at the interval that is n's highest divisor.

The reason Gun A is designed to fire spaceships in only odd numbered positions, and not odd and even numbered positions, is to keep it as simple as possible. Above 2 all primes are odd, so there is no need to construct even numbered spaceships only for them to be destroyed. The initial spaceship in position 2 is presented as a free gift at the beginning of the stream.

(i) Generation 0

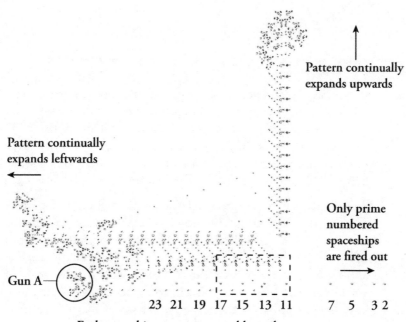

Pattern continually
expands upwards

Pattern continually
expands leftwards

Only prime
numbered
spaceships
are fired out

Gun A

23 21 19 17 15 13 11   7 5   3 2

Each spaceship represents an odd number

(ii) Generation 650

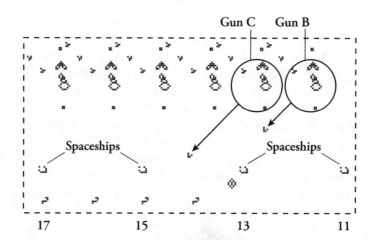

Gun C   Gun B

Spaceships   Spaceships

17   15   13   11

(iii) Detail from generation 650

Assumptions, Clarifications, References and Notes

CHAPTER ONE

p. 1 **As soon as** Every whole number divides into prime numbers in a unique way. For example, 2763 breaks down into 3 × 3 × 307, and this combination of prime numbers is the only one that equals 2763 when the numbers are multiplied together. The statement that every number can be broken down into a unique combination of prime divisors is also known as the Fundamental Theorem of Arithmetic.

2 **He paused. 'A new** The most celebrated arithmetical performance by a person with savant syndrome (the term used to describe someone with an autistic spectrum disorder who has prodigious skills) also involves prime numbers. In *The Man Who Mistook His Wife for a Hat* (Duckworth, 1985), Oliver Sacks tells the story of American twins John and Michael, who played a game in which they thought up six-figure numbers. He writes that they mulled over and savoured the numbers 'like two connoisseurs wine-tasting'. When Sacks checks the numbers, he realizes they are all primes, so he ups the ante by suggesting an eight-digit prime. This inspires the twins to swap bigger and bigger primes, including, an hour later, ones containing 20 digits, although by then Sacks has no way of checking if they are indeed primes. John and Michael, the 'prime twins', are not to be confused with 'twin primes', which are prime numbers, like 29 and 31, spaced 2 apart.

Allan W. Snyder, of the University of Sydney, believes that all humans have the mental machinery to perform savant-like calculations, but that this machinery cannot normally be accessed because of the way the brain is wired. In experiments, Snyder has shown that people's mathematical reasoning can improve if their brains are zapped by a small electrical current, a technique called 'transcranial direct current stimulation'. His theory is that the current temporarily inhibits neural activity, unlocking the savant in us all. While Snyder's research is controversial, similar results are being found at other universities.

2 **The word for one** Georges Ifrah, *The Universal History of Numbers*, John Wiley & Sons, 2000.

3 **One signifies unity** Vincent F. Hopper, *Medieval Number Symbolism*, Columbia University Press, 1938.

3 **Shakespeare is also** As well as the word 'odd', mathematics also provided another commonly used word for strange: 'eccentric'. This originally meant a circle around the Earth whose centre is not the Earth.

3 **Originally, the word** The word 'odd' comes from the Scandinavian *oddr*, for spear. From the shape of the spear came the Old Icelandic *oddi*, a triangle, or a tongue of land. (Oddi is also the name of the church school in south Iceland where Snorri Sturluson, the country's great poet and historian, lived in the twelfth century, and which is now a tourist destination.) The triangle led to the meaning of 'odd' as an unpaired member of a group of three, and then to the unpaired member of any group. (Source: the *Oxford English Dictionary*, and Anatoly Liberman, Oxford University Press, blog. oup.com/category/language-words/oxford_etymologist/)

4 **Again, we see** Yutaka Nishiyama, 'Odd and Even Number Cultures', *Mathematics for Scientists*, 2005.

4 **The Japanese taste** Yutaka Nishiyama, 'Why ¥2000 notes are unpopular', *Osaka Keidai Ronshu*, vol. 62, No. 5, 2012.

4 **US records show** Lee C. Simmons and Robert M. Schindler, 'Cultural Superstitions and the Price Endings Used in Chinese Advertising', *Journal of International Marketing*, 2003.

5 **In one experiment** Terence M. Hines, 'An odd effect: Lengthened reaction times for judgements about odd digits', *Memory & Cognition*, 1990.

5 **They showed respondents** James E. B. Wilkie and Galen V. Bodenhausen, 'Are numbers gendered?', *Journal of Experimental Psychology: General*, 2012.

6 **Respondents were about** Subsequent research by James E. B. Wilkie, as yet unpublished, shows that women tend to perceive number associations more strongly than men.

6 **The influential German** Vincent F. Hopper, *Medieval Number Symbolism*, Columbia University Press, 1938.

8 **In 2011, Dan** Dan King and Chris Janiszewski, 'The Sources and Consequences of the Fluent Processing of Numbers', *Journal of Marketing Research*, 2011.

10 **Manoj Thomas, a** Manoj Thomas, Daniel H. Simon, and Vrinda Kadiyali, 'The Price Precision Effect: Evidence from Laboratory and Market Data', *Marketing Science*, 2010.

11 **In 2008, researchers** Nicolas Guéguen et al., 'Nine-ending prices and consumers' behavior: A field study in a restaurant', *International Journal of Hospitality Management*, 2009.

11 **A price ending** William Poundstone, *Priceless*, Oneworld, 2010.

12 **For example, a** Sybil S. Yang, Sheryl E. Kimes, and Mauro M. Sessarego, '$ or Dollars: Effects of Menu-price Formats on Restaurant Checks', *Cornell Hospitality Report*, 2009.

12 **Another clever menu** In restaurants, the most common example of a column of numbers encouraging a purchase based on price rather than on the product is the tendency of customers to buy the second cheapest bottle of wine in the list. Buying the cheapest looks too miserly, especially if the meal is a romantic date. Restaurants often have their best mark-up on the second cheapest bottle.

12 **In one study** Birte Englich, Thomas Mussweiler and Fritz Strack, 'Playing Dice With Criminal Sentences: The Influence of Irrelevant Anchors on Experts' Judicial Decision Making', *Personality and Social Psychology Bulletin*, 2006.

13 **I had not realized** My web survey favouritenumber.net went live in 2011. When you clicked through from the title page, there was an introductory text followed by two sentences: 'My favourite number is …' and 'My reasons are …'. Respondents could answer in words or digits. The results in these pages are based on the first 33,516 responses, of which 3491 were null or void. When this book went to press, the total number of respondents was more than 42,000.

17 **In the middle of** Eviatar Zerubavel, *The Seven Day Circle*, Free Press, 1985.

18 **The Egyptians used** Georges Ifrah, *The Universal History of Numbers*, John Wiley & Sons, 2000.

18 **Psychologists have studied** Michael Kubovy and Joseph Psotka, 'The predominance of seven and the apparent spontaneity of numerical choices', *Journal of Experimental Psychology: Human Perception and Performance*, 1976.

18 **Such is the subconscious** There are only eight choices for two-digit odd numbers where both digits are different between 1 and

50, and the number 15 is mentioned in the set-up, so is unlikely to be suggested by the respondent. In *The Psychology of the Psychic* (Prometheus Books, 1980), authors David Marks and Richard Kammann tried the trick on a class of psychology students, more than a third of whom chose 37. The results were: 37 (35 per cent), 35 (23), 17 (10), 39 (10), 19 (9), 31 (5), 13 (5), others (3).

20 **Dan King and** Dan King and Chris Janiszewski, 'The Sources and Consequences of the Fluent Processing of Numbers', *Journal of Marketing Research*, 2011.

21 **In a similar** Marisca Milikowski, 'Knowledge of numbers: A study of the psychological representation of the numbers 1–100', PhD thesis at the University of Amsterdam, 1995.

CHAPTER TWO

27 **Curious to see** *Domesday Book: A Complete Translation*, Penguin Classics, 2003.

29 **The curious profusion** Simon Newcomb, 'Note on the Frequency of Use of the Different Digits in Natural Numbers', *American Journal of Mathematics*, 1881.

30 **More than half** Frank Benford, 'The law of anomalous numbers', *Proceedings of the American Philosophical Society*, 1938.

31 *Many real-world data*

Number	1	2	3	4	5	6	7	8	9
Benford's	30.1	17.6	12.5	9.7	7.9	6.7	5.8	5.1	4.6
Counties	30.2	18.8	12.2	9.9	7.1	6.3	5.7	4.8	5.0
Revenues	30.2	17.7	12.5	9.8	7.9	6.7	5.7	5.1	4.5

The population data is from the American Community Survey 2007–2011. The financial data consists of about 1.4 million data points, mined from Compustat by Jialan Wang.

32 **The method of** Scott de Marchi and James T. Hamilton, 'Assessing the accuracy of self-reported data: an evaluation of the toxics release inventory', *Journal of Risk and Uncertainty*, 2006; Walter R. Mebane Jr., 'Fraud in the 2009 Presidential Election in Iran?', *Chance*, 2010; Malcolm Sambridge et al., 'Benford's law in the natural sciences', *Geophysical Research Letters*, 2010.

39 **In the 1940s** Miles L. Hanley, *Word Index to James Joyce's Ulysses*, University of Wisconsin Press, 1953.

39 **The data was of** George Kingsley Zipf, *Human Behavior and the Principle of Least Effort*, Addison-Wesley, 1949.

41 **In fact, in all** The term for a word that only appears once in a text is *hapax legomenon*, which sounds like a character from an Asterix story, or a Scandinavian death metal band, and in this text appears only once.

44 **According to Richard** Richard Koch, *The 80/20 Principle*, Crown Business, 1998.

45 **For example, a Swedish** Fredrik Liljeros et al., 'The web of human sexual contacts', *Nature*, 2001.

45 **Researchers studying violence** N. Johnson et al., 'From old wars to new wars and global terrorism', arXiv:physics/0506213, 2005.

46 **Charles Darwin sent** João Gama Oliveira and Albert-László Barabási, 'Human dynamics: Darwin and Einstein correspondence patterns', *Nature*, 2005.

46 **Japanese academics looked** Takashi Iba et al., 'Power-Law Distribution in Japanese Book Sales Market', *Fourth Joint Japan-North America Mathematical Sociology Conference*, 2008.

47 **The magnitude of** Mark Buchanan, *Ubiquity*, Weidenfeld & Nicolson, 2000.

48 **Barabási's current field** Albert-László Barabási, *Linked*, Perseus, 2002; Albert-László Barabási, *Bursts*, Penguin, 2010.

49 **In recent years** Michael P. H. Stumpf and Mason A. Porter, 'Critical Truths About Power Laws', *Science*, 2012; Aaron Clauset, Cosma Rohilla Shalizi, and M. E. J. Newman, 'Power-Law Distributions in Empirical Data', *SIAM Review*, 2009.

50 **The bigger an animal** In his 1638 book *Discourses and Mathematical Demonstrations Relating to Two New Sciences*, Galileo sketched the following image of two bones, a small thin one and a large fat one. He wrote that for a large animal the big one would 'perform the same function which the small bone performs for its small animal'.

Nylabone, a pet accessories company, sells a nylon bone shaped like the fat one shown below. The Galileo, as it's called, is allegedly the 'world's strongest dog bone'.

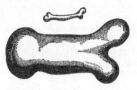

52 **In the 1930s** Melanie Mitchell, *Complexity: A Guided Tour*, Oxford University Press, 2009.

52 **The physicist Geoffrey** Geoffrey B. West, James H. Brown, and Brian J. Enquist, 'A General Model for the Origin of Allometric Scaling Laws in Biology', *Science*, 1997.

53 **West and his** Luís M. A. Bettencourt et al., 'Growth, innovation, scaling, and the pace of life in cities', *PNAS*, 2007.

CHAPTER THREE

57 **Rob, who is** Rob knows of 6177 trig pillars in Great Britain, including 45 that are just debris and 100 that have been toppled. Most of the ones he has yet to see are on islands. He has seen two pillars on Ministry of Defence land from a distance away, including one located in the nuclear submarine base in Coulport, Scotland, but he has not been allowed to approach them. Only four other baggers have passed 3000 pillars.

58 **The distance along the** The practicalities of Thales's pyramid measurement are discussed in 'Thales' Shadow' by Lothar Redlin, Ngo Viet and Saleem Watson, in *Mathematics Magazine*, 2000. The authors show that the sun is only perpendicular to the pyramid twice a day during spring and summer: once in the morning and once in the evening.

59 **By deducing a** It may be the case that the Egyptians knew more mathematics than they are given credit for, but it is impossible to know since so little has survived.

59 **Thales is also** Carl B. Boyer, *A History of Mathematics*, John Wiley & Sons, 1968.

62 *For right-angled triangles* The 'other two sides' of a right-angled triangle are historically called the *catheti* (singular: *cathetus*), although this word is obsolete in English. In other languages, however, the word survives: *Kathete* in German, and *cateto* in Spanish and Portuguese.

63 **Greek numerical notation** Florian Cajori, *A History of Mathematical Notations*, Dover, 1993.

63 **Each power of** Georges Ifrah, *The Universal History of Numbers*, John Wiley & Sons, 2000.

64 **The shrinking values** The most versatile system that uses only unit fractions is a binary system, in which the fractions are made up of halves, halves of halves, halves of halves of halves, and so on. Or, $\frac{1}{2}, \frac{1}{4}, \frac{1}{8}, \frac{1}{16}$ ... Using this system every possible fraction can be described uniquely in terms of unit fractions. In 1911, the Egyptologist Georg Möller wrote that he had discovered in his research a wonderfully picturesque ancient notation for the first six binary unit fractions. In the Eye of Horus, illustrated below, each element represents a quantity: the left cornea is $\frac{1}{2}$, the iris $\frac{1}{4}$, the eyebrow $\frac{1}{8}$, and so on, with other parts representing $\frac{1}{16}, \frac{1}{32}$ and $\frac{1}{64}$. The 63 possible non-zero assemblies of pieces in the Eye of Horus can be used to express every fraction from $\frac{1}{64}$ to $\frac{63}{64}$. As well as being a sexy image, it had a sexy story: the eye is the mystical symbol of the falcon-god Horus, whose eye had been ripped into six pieces by his uncle and then reassembled. Unfortunately, after almost a century of acceptance, in 2002 the Eye of Horus was debunked by Jim Ritter – there is no evidence that there ever was such a thing. Jim Ritter, 'Closing the Eye of Horus: the Rise and Fall of "Horus-Eye Fractions"', *Under One Sky: Astronomy and Mathematics in the ancient Near East*, 2002.

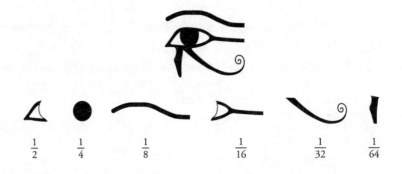

$\frac{1}{2}$ $\frac{1}{4}$ $\frac{1}{8}$ $\frac{1}{16}$ $\frac{1}{32}$ $\frac{1}{64}$

64 **By the time** The Greek system in full was:

α	β	γ	δ	ε	ς	ζ	η	θ
1	2	3	4	5	6	7	8	9

ι	κ	λ	μ	ν	ξ	ο	π	ϟ
10	20	30	40	50	60	70	80	90

ρ	σ	τ	υ	φ	χ	ψ	ω	ϡ
100	200	300	400	500	600	700	800	900

74 **In 1533 the Dutch** Eli Maor, *Trigonometric Delights*, Princeton University Press, 1998.

75 **For Britain, on** John Keay, *The Great Arc*, HarperCollins, 2000.

CHAPTER FOUR

81 **When you bounce** Assuming no spin on the ball.

83 **The giant dome of** lds.org/locations/temple-square-salt-lake-city-tabernacle

83 **Ancient Greek mathematics lasted** Carl B. Boyer, *A History of Mathematics*, John Wiley & Sons, 1968.

83 **The ellipse comes** In addition to its mathematical sense, the Latin *parabola* also meant parable – since an alternative meaning for 'to throw alongside' is 'to compare'. A parable is a simple story that is compared with a complicated one. From this meaning came the French 'parler' and many English words, from 'parliament' to 'parole'.

85 **In its final** Arthur Koestler, *The Sleepwalkers*, Hutchinson, 1959.

85 **Remarkably, cycles and epicycles** The mathematical reason why cycles and epicycles can describe any closed and continuous orbit is explained by two concepts I deal with elsewhere in this book: complex numbers and Fourier series. In the same way that a wave can be broken down into sinusoids, a path in the complex plane can be broken down into a combination of circular rotations.

86 **In 2005 the** Santiago Ginnobili and Christián C. Carman, 'Deferentes, Epiciclos y Adaptaciones', *Filosofia e história da ciência no Cone Sul*, 2008.

87 **At 4.37am** Arthur Koestler, *The Sleepwalkers*, Hutchinson, 1959.

89 **The philosopher Norwood** Norwood Russell Hanson, *Patterns of Discovery*, CUP, 1961. Hanson started off as a trumpet player, before

becoming a fighter pilot during the Second World War. Known as the Flying Professor, he continued flying planes in peacetime, and was famous for performing aerobatics. He died aged 42 when his plane crashed in dense fog in New York state.

90 **Galileo accepted Copernicus's** David Wootton, *Galileo, Watcher of the Skies*, Yale University Press, 2010.

91 **From Galileo's knowledge** Stillman Drake and James MacLachlan, 'Galileo's Discovery of the Parabolic Trajectory', *Scientific American*, 1975.

96 **Cartesian coordinates determine** Descartes generally used oblique axes, so the modern understanding of 'Cartesian' coordinates on perpendicular axes is due to others who clarified his work.

98 **In the mid-nineteenth** A. F. Möbius, 'Geometrische Eigenschaften einer Factorentafel', *Journal für die reine und angewandte Mathematik*, 1841.

98 **He never mentioned it** Rodolphe Soreau, *Nomographie; ou, Traité des abaques*, Chiron, 1921; Ron Doerfler, 'The Lost Art of Nomography', *The UMAP Journal*, 2009; H. A. Evesham, 'Origins and Development of Nomography', *Annals of the History of Computing*, 1986.

100 **The hyperbola stands** Martin Gardner, *Mathematical Games: The Entire Collection of His Scientific American Columns*, CD, 2005.

102 **In the seventeenth** J. A. Bennett, *The Mathematical Science of Christopher Wren*, CUP, 1982.

102 **In the nineteenth** *Programme for Intersections: Henry Moore and Stringed Surfaces*, exhibition at the Royal Society, 2012.

CHAPTER FIVE

109 **Or so wrote** Bob Palais, 'π Is Wrong!', *The Mathematical Intelligencer*, 2001.

109 **Many agree with** Among those historical figures who preferred the ratio $\frac{\text{circumference}}{\text{radius}}$ was Al-Kashi, who in fifteenth century Samarkand is said to have calculated pi to 14 decimal places, a more accurate result than anyone before him. In fact, Al-Kashi did not calculate pi at all, but calculated the ratio $\frac{\text{circumference}}{\text{radius}}$ to 14 decimal places. In 1698 Abraham de Moivre used the symbol $\frac{c}{r}$ for $\frac{\text{circumference}}{\text{radius}}$, but it didn't catch on.

110 **In the Tau Manifesto** tauday.com

110 **The symbol τ** I'd even go so far as to say it is quadruply clever. It also works as a homage to Terence Tao, the Australian Fields Medallist, who is a professor at UCLA.

112 **Never in the** John Martin, 'The Helen of Geometry', *The College Mathematics Journal*, 2010; E. A. Whitman, 'Some Historical Notes on the Cycloid', *American Mathematical Monthly*, 1943; Martin Gardner, *Mathematical Games: The Entire Collection of His Scientific American Columns*, CD, 2005.

114 **Inspired by his** Huygens built some pendulums with cycloidal cheeks, but because of problems like friction they didn't work in practice any better than a simple, cheekless pendulum. His solution was to use a simple pendulum but to swing it only slightly, since for very small amplitudes a bob will take approximately the same time to complete a full oscillation.

119 **One of my favourite** Yes, I do have other favourite pub maths puzzles, such as this one also involving coins. Take six coins and position them as shown below left. The challenge is to rearrange the coins into a hexagon by sliding them one by one. A coin can only be slid to a position where it is touching two other coins. You are not allowed to lift the coin off the table, nor slide it over another coin, nor move other coins out of the way. Can you rearrange the coins in three slides?

If you managed that one, now try to rearrange a triangle into a line in seven slides, again with the rule that a coin can only be slid to a position where it touches two other coins.

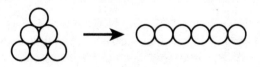

The next time you're waiting in a bar, and have some loose change, try it!

121 **The first person** Roberval's sinusoid appears in his drawing of how to find the area under a cycloid. It is unlikely he was aware that the curve had anything to do with the trigonometric ratio sine.

121 **The vertical position** Assuming no loss of energy through friction.

123 **The Victorian harmonograph** Robert J. Whitaker, 'Harmonographs. I. Pendulum design', *American Journal of Physics*, 2001; Robert J. Whitaker, 'Harmonographs. II. Circular design', *American Journal of Physics*, 2001.

124 **Around the time that** Eli Maor, *Trigonometric Delights*, Princeton University Press, 1998.

128 **In 1798 Joseph** John Herivel, *Joseph Fourier: The Man and the Physicist*, Clarendon Press, 1975.

128 **It is said** I. B. Cohen, *The Triumph of Numbers*, W. W. Norton & Company, 2005.

130 **The sum of** The Fourier series of a wave is written as the following formula:

$$k + a_1 \sin x + a_2 \sin 2x + a_3 \sin 3x + a_4 \sin 4x + \ldots$$
$$+ b_1 \cos x + b_2 \cos 2x + b_3 \cos 3x + b_4 \cos 4x + \ldots$$

where k is a constant, and the a's and b's are the amplitudes of the corresponding sinusoids.

CHAPTER SIX

135 **In Boulder, Colorado** youtube.com/watch?v=F-QA2rkpBSY

135 **Albert Bartlett, emeritus** Albert Bartlett died aged 90 on 7 September, 2013. He had given his lecture 1742 times.

139 **In 1980 the** Gideon Keren, 'Cultural differences in the misperception of exponential growth', *Perception & Psychophysics*, 1983.

140 **In 1973 Daniel** Daniel Kahneman and Amos Tversky, 'Availability: A heuristic for judging frequency and probability', *Cognitive Psychology*, 1973.

147 **Bernoulli discovered** *e* Eli Maor, *e: The Story of a Number*, Princeton University Press, 1994.

151 **He was also** J. E. Hofmann, from the biography of Jakob Bernoulli in the *Dictionary of Scientific Biography*, Scribner, 1970.

151 **Johann relished solving** William Dunham, *Journey Through Genius*, Penguin, 1991.

153 **The arch, in fact,** Santiago Huerta, 'Structural Design in the Work of Gaudí', *Architectural Science Review*, 2006.

160 **Euler started at the** Ed Sandifer, 'How Euler Did It', *MAA Online*, 2004.

161 **After the German** Arthur Koestler, *The Sleepwalkers*, Hutchinson, 1959.

162 **Consider the following** Martin Gardner, *Mathematical Games: The Entire Collection of His Scientific American Columns*, CD, 2005.

163 **The Secretary/Marriage Problem** Many variants of the problem have been studied, such as how the probabilities change if the interviewer can recall certain applicants, or how hiring an early applicant could save costs. Darren Glass examined 'The Secretary Problem from the Applicant's Point of View', in *The College Mathematics Journal*, 2012. If the number of applicants is at least nine, then the optimal interview position is last. But there is no margin for error. 'Being the final interviewee gives one the highest probability of being hired, but being the second-to-last has one of the lowest!' he wrote. 'Students going on the job market should put their energy into improving their résumés rather than strategizing interview timing.'

164 **Boris Berezovsky, the** Boris Berezovsky died aged 67 on 29 March 2013.

164 **Here is a game** Theodore Hill, 'Knowing When to Stop', *American Scientist*, 2009.

CHAPTER SEVEN

169 **'They fobbed me** '"Cool Cash" card confusion', *Manchester Evening News*, 2007.

170 **In ancient Asia** Georges Ifrah, *The Universal History of Numbers*, John Wiley & Sons, 2000.

171 **The following argument** Martin Gardner, *Mathematical Games: The Entire Collection of His Scientific American Columns*, CD, 2005.

171 **In the fifteenth** An absurd number is not to be confused with a 'surd', which is a synonym for an irrational number – that is, a number that cannot be expressed as a ratio of two whole numbers. The Greeks used the word *alogos*, meaning 'not a ratio', for the irrational numbers. But the word also means 'not speaking', which the Arabs translated as *assam*, or deaf. Latin texts used the word *surdus*, the direct translation from the Arabic, so irrational numbers became 'deaf' numbers, or surds.

173 **The German philosopher** Alberto A. Martínez, *Negative Math*, Princeton University Press, 2006.

173 **Yet even at the end** William Frend, *The Principles of Algebra*, G. G. and J. Robinson, 1796. The book allowed minus signs, but forbade unknown quantities – which could denote real things – from taking negative values.

Frend is best remembered as a social reformer and radical. He was banished from Cambridge after a high profile trial, for denouncing the Church of England. His supporters included Samuel Taylor Coleridge. Frend's daughter Sophia (who married the eminent mathematician Augustus de Morgan) wrote that her father's 'mental clearness and directness may have caused his mathematical heresy, the rejection of the use of negative quantities in algebraical operations', adding that 'it is probable that he thus deprived himself of an instrument of work, the use of which might have led him to greater eminence in the higher branches.'

176 **The first person to consider** Paul J. Nahin, *An Imaginary Tale*, Princeton University Press, 1998.

177 **Euler gave the** Euler was the first person to use i to mean $\sqrt{-1}$, but he only used the symbol once, in a memoir published 11 years after he died. It was only after Gauss adopted i in 1801 that others began to make systematic use of it.

178 **Amazingly, the solution** Another solution to $x^2 = i$ is $-\frac{1}{\sqrt{2}} - (\frac{1}{\sqrt{2}})i$, which is the negative of the solution in the text.

178 **The nature of the** Ed Leibowitz, 'The Accidental Ecoterrorist', *Los Angeles Magazine*, 2005.

180 **A radian is** James Thomson, who appears in Chapter Eight, coined the term 'radian' in 1873, even though by then the concept was a century and a half old.

185 **In fact, the Schrödinger** The Schrödinger wave equation is:

$$i\hbar \frac{\partial}{\partial t} \Psi = \hat{H}\Psi$$

Where i is $\sqrt{-1}$, $\hbar$ is the reduced Planck constant, Ψ is the wave function of the quantum system, and $\hat{H}$ is the Hamiltonian operator.

189 **Hamilton's peers ridiculed** Melanie Bayley, 'Algebra in Wonderland', *The New York Times*, 2010.

190 **One of the main** John C. Baez and John Huerta, 'The Strangest Numbers in String Theory', *Scientific American*, 2011.

190 **Bertrand Russell, the only** Bertrand Russell, 'The Study of Mathematics', *Mysticism and Logic: And Other Essays*, Longman, 1919. Bertrand Russell is the only *world-class* mathematician to have won the Nobel Prize for Literature. But both Alexandr Solzhenitsyn (who won the Nobel Prize in 1970) and J. M. Coetzee (2003) have undergraduate degrees in maths.

199 **In 1991 the** Dave Boll did not publish in a journal, but in a post on a forum about fractals: groups.google.com/forum/?hl=en#!topic/sci.math/jHYDf-Tm0-8_

CHAPTER EIGHT

204 **The maths community** In 2001 the Norwegian government established the annual Abel Prize, named after its countryman the mathematician Niels Henrik Abel (1802–1829). It's worth approximately $1 million. While being similar to the Nobel in monetary value and Scandinavianness, it does not yet have the prestige of the Fields.

204 **The British mathematician** gowers.wordpress.com

205 **Yet the historian Plutarch** Plutarch, *Life of Marcellus*, as quoted in the MacTutor History of Mathematics online archive.

206 **Archimedes first drew** Carl B. Boyer, *The History of the Calculus and Its Conceptual Development*, Dover, 1959.

The large triangle is constructed such that the tangent at the bottom vertex of the triangle is parallel to the original line. Likewise, each time a new triangle is constructed, the new vertex is located such that the tangent at that vertex is parallel to the opposite side.

209 **He started thinking about** Ernst Sondheimer and Alan Rogerson, *Numbers and Infinity*, Dover, 2006.

211 **In 1665, Isaac** James Gleick, *Isaac Newton*, Harper Perennial, 2003.

211 **A body that moves** Ian Stewart, *17 Equations that Changed the World*, Profile Books, 2012; Charles Seife, *Zero*, Souvenir Press, 2000.

218 **For more than two** A. Rupert Hall, *Philosophers at War*, Cambridge University Press, 2002.

219 **Only when the** Augustus De Morgan, *A Budget of Paradoxes*, 1872.

221 **The equation describes** $f(t, x, v)$ is a probability density function that gives the probability of particles having a position near x and a speed near v at time t. The symbol ∇ means gradient, but applied to several variables. Cédric Villani, *Théorème vivant*, Grasset, 2012.

226 **'After half a century** *The Railroad Gazette* (now *Railway Age*), 1880, quoted in Halsey G. Brown, 'The History of the Derivation of the AREMA Spiral', arema.org

228 **Holbrook's idea for** The clothoid is the curve such that curvature is proportional to length. Written algebraically, curvature = ks, where k is an arbitrary constant and s is the distance along the curve from the origin. The Belgian mathematician Franki Dillen devised a whole new class of spirals by letting the curvature be a polynomial term in s. (A polynomial is an expression constructed of variables, and powers of variables, using only addition, subtraction and multiplication.) He called them 'polynomial spirals', and they are very pretty. The 'Picasso spiral' was one of his favourites:
Curvature = $10 \, (-45 + 51s - 18s^2 + 2s^3)$.

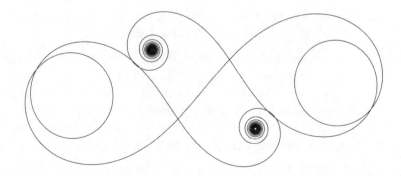

228 **When, in the** Joe Moran, *On Roads*, Profile Books, 2009.

229 **In the mid-nineteenth** Robert Cartmell, *The Incredible Scream Machine*, Amusement Park Books, 1987; 'Chemin de Fer Aérien', *La Nature*, 1903.

Before M. Clavières's loop-the-loop was opened to the public, three tests were made: the first with monkeys as passengers, the second with weights heavier than a large human, and finally with an acrobat.

231 **'What are these** George Berkeley, *The Analyst: Or, a Discourse Addressed to an Infidel Mathematician*, 1734.

CHAPTER NINE

234 **Mathematical proof is** Steven G. Krantz, *The Proof is in the Pudding*, Springer, 2011.

235 **If you spill** Martin Gardner, *Mathematical Games: The Entire Collection of His Scientific American Columns*, CD, 2005.

236 **Between the First** Lwów is now Lviv, Ukraine.

238 **Theorem:** *All numbers* At time of writing the best candidates for lowest boring number are 224, which is the lowest number not to have its own page on Wikipedia, and 14,228, which is the lowest number not to appear in the *Online Encyclopedia of Integer Sequences*. But by being written about here, of course, the numbers are now interesting.

242 **There is a pattern** If n is the number of dots, then the number of regions is the equation $\frac{1}{24}\,(n^4 - 6n^3 + 23n^2 - 18n + 24)$.

246 **His idea was** To some philosophers of maths, although not to Frege, the statement 'the negation of the negation of statement A implies statement A' was deeply controversial.

246 **Some of my favourite** Douglas R. Hofstadter, *Metamagical Themas*, Basic Books, 1996.

248 **The Polish logician** Martin Gardner, 'Logical Paradoxes', *The Antioch Review*, 1963.

248 **Comedians, as well** John Allen Paulos, *I Think, Therefore I Laugh*, Penguin, 2000.

248 **Axiomatic set theory** One of the principal aims of set theory was

to show that mathematics was 'complete', meaning that if a theorem is true, then it is provable within the system. Yet in 1931 the Austrian Kurt Gödel proved that this is not the case: in any system powerful enough to include arithmetic, there will be some true statements that are neither provable nor disprovable. Gödel's work had a great impact on mathematical philosophy, as it limited the scope of logic as a foundation for mathematics.

250 **'Whereas in the past** Nicolas Bourbaki, *Theory of Sets*, Hermann, 1968.

It is curious, at the very least, that Bourbaki makes no mention of Kurt Gödel. (See previous note.)

250 **Nicolas Bourbaki, they** Poldavia, or *Poldévie*, was a joke country invented in 1929 by a right-wing French journalist in a letter to left-wing members of parliament asking them to intervene on behalf of its oppressed people. After Bourbaki adopted the country as a homeland, it became a serial joke in the work of several French writers of the post-war period. According to David Bellos, professor of French at Princeton and the author's dad, it is 'a rare example of mathematical humour turning into a literary motif'.

250 **During the first of** Maurice Mashaal, *Bourbaki*, American Mathematical Society, 2006.

251 **In 1999, the** A. R. D. Mathias, 'A term of length 4,523,659,424,929', *Synthese*, 2002.

252 **Western governments realized** Bob Moon, 'Who Controls the Curriculum? The story of "New Maths" 1960–1980', *International Perspectives in Curriculum History*, 1987.

254 **A 'proof assistant'** There are currently more than a dozen proof assistants, the best-known being *Coq*, *HOL Light*, *Isabelle* and *Mizar*, which was started in Poland in the 1970s and which its users claim has the largest coherent body of formalized proofs.

254 **The aspiration is** That is, to prove all of mathematics that is provable. (See previous note on Gödel.)

254 **Mathematicians responded uneasily** Steven G. Krantz, *The Proof is in the Pudding*, Springer, 2011.

CHAPTER TEN

258 **No other lines**

> **Theorem:** When sieving up to n you only need to count the primes up to $\sqrt{n}$.

> **Proof:** Imagine you have sieved up to $\sqrt{n}$ but there is still a non-prime number m between $\sqrt{n}$ and n. The number m is non-prime, which means it must have prime factors. Since we have sieved all primes less than $\sqrt{n}$ the prime factors of m must be larger than $\sqrt{n}$. But the multiplication of two or more numbers larger than $\sqrt{n}$ is bigger than n, so there can be no m less than n. QED.

262 **There were so many** One of his favourite names without a game was *Don't Ring Us, We'll Ring You.*

266 **In 1970 the** Martin Gardner, *Mathematical Games: The Entire Collection of His Scientific American Columns*, CD, 2005.

266 **With computers, the** For anyone wanting to play the Game of Life, and I recommend it, the best software is Golly, downloadable from golly.sourceforge.net

266 **At MIT, the** Steven Levy, *Hackers*, O'Reilly Media, 2010.

268 **The Life sieve** The first sieve of Eratosthenes was designed in 1991 by Dean Hickerson. The sieve mentioned here is an improved version designed by Jason Summers in 2005.

269 **Conway constructed the** William Poundstone, *The Recursive Universe*, Oxford University Press, 2005.

275 **Gemini has heightened** As this book was going to press, the American Lifenthusiast Dave Greene announced a stunning new self-replicating pattern that reduces the number of live cells required for a constructor unit from Gemini's 16,229 to only 256. He called his pattern a 'Geminoid' replicator because it uses some of Gemini's technology. Unlike Gemini, however, it has only one constructor unit, which survives after replication, rather than two that are destroyed once they have served their purpose. Dave's Geminoid replicator gives birth to an identical copy, which gives birth to an identical copy, and so on ad infinitum, creating a line of descendants that spreads across the grid. Since the constructor is made up of such a small number of live cells, it

makes building new patterns easier. Dave hopes that Geminoid technology will lead to many new types of replicating objects.

280 **His opinions are** Stephen Wolfram, *A New Kind of Science*, Wolfram Media, 2002.

INTERNET

Mathematics is very well served on the web. I made ample use of Wolfram MathWorld, the MacTutor History of Mathematics, and, inevitably, Wikipedia.

Acknowledgements

At Bloomsbury in London, thanks to Bill Swainson, Alison Glossop, Laura Brooke, Helen Flood, Amanda Shipp, Greg Heinimann, David Mann, Richard Atkinson, and especially Xa Shaw Stewart for meticulously overseeing every fraction and exponent. At Simon & Schuster in New York: Ben Loehnen, Emily Loose and Brit Hvide, and at Doubleday in Toronto: Tim Rostron.

Ben Sumner was a demon copyeditor, and Edmund Harriss, Yin-Fung Au, June Barrow-Green, Erica Jarnes and Gareth Roberts provided invaluable comments on the text. Thanks also to Simon Lindo for the illustrations, The Surreal McCoy for the cartoons and Susan Wightman at Libanus Press for the design and typesetting.

I relied on interviews and correspondences with many people, and I am sincerely grateful for the time they gave me.

Chapter One: Jerry Newport, Greg Rowland, Manoj Thomas, Terence Hines, Jim Wilkie, Husam E. Sadig, Saffi Haines, Dan King, Tom Dearden, Jer Thorp, Francesca Stavrakopoulou, Francesca Rochberg, Richard Wiseman, David Marks, Sophie Scott, Stephen Macknik, Peter Lynn, Yutaka Nishiyama, Robert Schindler.

Chaper Two: Will Rennie, Jialan Wang, Ted Hill, Erika Rogers, Darrell Dorrell, Albert-László Barabási, David Hand, Walter Mebane, Christiane Fellbaum, Jure Leskovec, Geoffrey B. West, Pete Whitelock.

Chapter Three: Michalis Sialaros, Apostolos Doxiadis, Mark Greaves, Robert Woodall, John Keay, Darran Shepherd.

Chapter Four: Ramiro Serra, Ron Doerfler, Ian Dickerson, Silvia Pezzana, Art P. Frigo Jr.

Chapter Five: Bob Whitaker, Ivan Moscovich, Tom Armstrong, Brett Crockett, John Whitney Jr., Karl Sims.

Chapter Six: Andrew Smith, Roger Ridsdill Smith, Nikolai Malsch, Albert Bartlett, Tim Harford, Stan Wagon.

Chapter Seven: John Baez, David Tong, Dave Makin, Brian Pollock, Cliff Pickover, Daniel White, Orson Wang, Robert L. Devaney.

Chapter Eight: Peter Hopp, Bill Thacker, John Wardley, Werner

Stengel, Cédric Villani, Franki Dillen, Hartosh Bal.

Chapter Nine: Alex Paseau, Jim Holt, Norman Megill, Lawrence Paulson, Nathaniel Johnston.

Chapter Ten: Tom Rokicki, Adam P. Goucher, Tim Hutton, Paul Chapman, Dave Green, Adam Rutherford, Stefanie Prather, Jean Buck, Stephen Wolfram, Bill Gosper, Andy Adamatzky, Tim Hutton, Nick Gotts, John Conway, Craig Lent, Doug Tougaw.

I feel extremely fortunate to be represented by Janklow & Nesbit. Thanks to my agent Rebecca Carter and her colleagues Rebecca Folland, Kirsty Gordon, Lynn Nesbit and Claire Paterson.

Friends and family have generously provided help when called upon, from moral support to mathematical clarifications to the use of a Paris flat: Gavin Pretor-Pinney, Hugh Morison, Cliff Pickover, Graham Farmelo, James Grime, Colin Wright, Cordelia Jenkins, Francesca Segal, Roger Highfield and Simon Kuper. Thanks to my parents David Bellos and Ilona Morison for their unceasing encouragement and faith, and most of all to my wife Natalie for her many contributions to this book and the happiness of its author.

I mentioned my niece at the end of the acknowledgements of *Alex's Adventures in Numberland* in return for her doing well at her Maths A-level. She now gets a name check for her decision to study Mathematics and Psychology at university. Go Zara!

Picture Credits

p. 9 Courtesy of Dan King.

p. 76 Science Museum/Science & Society Picture Library.

p. 95 © Iain Frazer/Shutterstock.com.

p. 99 Taken from *Nomographie* by Rodolphe Soreau, Chiron, 1921.

p. 103 © Kletr/Shutterstock.com.

p. 117 © The British Library Board, 48.d.13.16, vol. 2, title page.

p. 118 © Alex Bellos.

p. 122 Taken from *Sound* by Alfred Marshall Mayer, Macmillan and Co., 1879.

p. 123 © Karl Sims, www.karlsims.com.

p. 125 Taken from *Sound* (Third Edition) by John Tyndall, Longmans, Green and Co., 1875.

p. 154 Photograph by Natalie Bellos.

p. 156 © Stan Wagon.

pp. 195–7 © Brian Pollock.

p. 229 From *L'Illustration*, 1846.

p. 230 © Werner Stengel.

p. 281 © iStock.com/busypix.

Index

Berezovsky, Boris, 164, 315
Berkeley, Bishop George, 231
Bernoulli, Daniel, 151, 156
Bernoulli, Jakob, 145–7, 150–1
Bernoulli, Johann, 115–17, 145,
    150–1, 156, 218
bicycle, square-wheeled, 155–6
Big Bang Theory, The, 14
bisector, 285
Blackburn, Hugh, 122
blue whale, 52
Bodenhausen, Galen, 5–6
Boll, Dave, 199
Boltzmann equation, 204–5,
    221–2, 318
Bombelli, Rafael, 176, 178
bones, thickness of, 50, 308–9
book sales, 46–7
Borsuk, Karol, 237
Boulder, population growth in,
    137–8
Bourbaki, Charles Denis, 250
Bourbaki mathematics, 250–4
Boyle's law, 104
brachistochrone, 115–16
Brahe, Tycho, 87
Brahmagupta, 171
brain, number processing in, 5
Brandram, Samuel, 28
Britain, national triangulation, 77
Brown, James, 52
Buddhist beliefs, 4
Buée, Abbé Adrien-Quentin, 183
Bush, Vanevar, 225

cakes, division of, 238, 300
calculus, 117, 205, 208, 211–19,
    224–5, 231, 244–5, 285
    Fundamental Theorem of,
    216–17, 287
cannonballs, 90, 157
Canton Tower, 104
Cantor, Georg, 245
card games, 159–61
Cardano, Girolamo, 176
cardioid, 118
Carman, Christián, 86
Carnot, Nicolas, 224
Carroll, Lewis (Charles Dodgson),
    189
catenary curve, 150–5
    and architecture, 152–5
cellular automata, 261–82, 285
    one-dimensional, 276–8
    quantum-dot, 282
    see also Game of Life
centripetal force, 225–7
Chapman, Paul, 257, 271–2,
    274–5
chess, 159, 241
Chinese Americans, 5
chords, 66, 285
    half-chords, 67–8, 120
circle
    and conic sections, 79–80
    curvature, 227–8
    definition of, 109–10
    division into degrees, 65, 180
    division into infinitesimal
    polygons, 209
    division into radians, 180
    division into regions, 242
    and ellipse, 81
circle constant, 285

A Note on the Author

Alex Bellos is the bestselling author of *Alex's Adventures in Numberland*, which was shortlisted for the BBC Samuel Johnson Prize. He is the *Guardian*'s maths blogger, and has worked for the paper in London and Rio de Janeiro, where he was its unusually numerate foreign correspondent. He is a curator-in-residence at the Science Museum and has a degree in Mathematics and Philosophy from the University of Oxford. He lives in London.

A Note on the Type

The text of this book is set Adobe Garamond. It is one of several versions of Garamond based on the designs of Claude Garamond. It is thought that Garamond based his font on Bembo, cut in 1495 by Francesco Griffo in collaboration with the Italian printer Aldus Manutius. Garamond types were first used in books printed in Paris around 1532. Many of the present-day versions of this type are based on the *Typi Academiae* of Jean Jannon cut in Sedan in 1615.

Claude Garamond was born in Paris in 1480. He learned how to cut type from his father and by the age of fifteen he was able to fashion steel punches the size of a pica with great precision. At the age of sixty he was commissioned by King Francis I to design a Greek alphabet, for this he was given the honourable title of royal type founder. He died in 1561.